Information and Communication Technology:

The Industrial Revolution That Wasn't

Bernard C. Beaudreau

Information and Communication Technology:
The Industrial Revolution That Wasn't

Lulu.com

ISBN: 978-1-4357-1772-5

Preface

In 1998, Robert Solow dared to say what many had been thinking for some time, namely that "we see computers everywhere but in output, growth and productivity data." In other words, computers had failed to generate the much anticipated increase in output growth. As it turns out, this was the first salvo. Where there was once hope and anticipation, there was now doubt, doubt over whether ICT would deliver the goods, so to speak. The doubting Toms, however, were few and far-between. After all, the ICT had opened up a brave new world, albeit a virtual one. Surely, the mother lode was not far behind. Scholars responded, defending for the most part, the ICT dream. According to Alvin Toffler, ICT would usher in the "Third Wave," of comparable or even greater magnitude to the first two.

The jury is still out. While doubt grows with every passing day, month and year, ICT continues to enjoy considerable support in the scholarly community as well as in the population in general. While naysayers are growing in number, most observers remain optimistic. Lacking from this debate, however, is a well-articulated model of the role of information in material processes, and as such in economic growth. This, as it turns out, is where this book got its start. The question of the role of ICT in material processes and economic growth provided a unique opportunity to apply the energy-organization approach to modeling material processes (Beaudreau 1998,1999) to a real world issue. Is the ICT revolution comparable to the first and second industrial revolutions? If so, on what bases; if not, why not?

This book is an attempt at imparting a dose of science into the debate. In the place of the heuristics and optimism that characterizes the current debate is a theoretical model of material processes that is sufficiently general to incorporate energy and information, as well as the standard factor inputs (capital and labor), into the analysis.

I would like to thank those who over the years provided useful comments on earlier versions of the manuscript. All remaining errors, however, are my sole responsibility.

Contents

Introduction

> You can see the computer age everywhere these days, except in the productivity statistics.
>
> —Robert Solow

> ...the productivity performance of the manufacturing sector of the United States economy since 1995 has been abysmal rather than admirable. Not only has productivity growth in non-durable manufacturing decelerated in 1995 to 99 compared to 1972 to 95, but productivity growth in durable manufacturing stripped of computers has decelerated even more.
>
> —Robert Gordon

For the past three decades, the West has lived in a constant state of anticipation, anticipation of a third economic coming of sorts, of a new economic era—in short, of a replay of the first and second industrial revolutions. The launching of new computers and operating systems (e.g. Windows XP, Vista) has conjured up and continues to conjure up visions of great wealth, of great potential, of a brave new world where everything or nearly everything is possible. With its ability to mimic every process, every product and every datum in virtual space, information and communication technology (ICT) has raised expectations in general. The possibilities appear infinite.

As if caught up in this swell of enthusiasm and excitement, the scientific community has been relatively mute on the validity of such claims. Can ICT do what the steam engine and the electromagnetic motor did for the 19th and 20th centuries respectively? Can ICT raise the standard of living the way these two power transmission technologies did for Great Britain and the United States and ultimately, the world? If so, then how? If not, then why not? What are the relevant fundamentals? The underlying laws?

In the late 1990's Nobel laureate Robert Solow was one of the first to openly question the received wisdom regarding ICT and productivity growth. As he put it, computers are everywhere, except in productivity data, a finding he dubbed the Information Paradox. If nothing else, a red flag had been raised. Paul David, in a now classic article entitled "The Dynamo and the Computer: An Historical Perspective on the Modern Productivity Paradox," published in the American Economic Review sought to allay growing fears that ICT was failing to live up to its billing by comparing it to the introduction of the dynamo a century ago, pointing to the presence of significant diffusion lags in the early 20th century, and extrapolating to the ICT case.[1] Robert Gordon on the other hand has gone so far as to argue that the two are not comparable, at least in so far as their effect on productivity.

Plaguing the literature on the ICT revolution has been what we choose to refer to as "weak fundamentals." Specifically, the role of information in general and ICT in particular in production is poorly developed, making for more confusion than insight. How ICT actually affects productivity is very much an open question, one that has yet to be answered satisfactorily. According to Dale Jorgenson, ICT enters the production function via the capital input, specifically via ICT-based capital.[2]

Left unanswered, however, are the basic questions: what is the exact role of information in production?; how does more or better information affect productivity?; and how has ICT affected material processes in general? Including it in the Cobb-Douglas—technology scaler is, in our view, no longer acceptable, especially in light of its equivocal nature.

This is where this work fits in, namely, to provide a scientifically-consistent (consilient as defined by Edward O. Wilson) view of the role of information in material processes and reexamine the many claims made by engineers, scientists, futurists, and growth theorists regarding ICT and the purported third industrial revolution. Using a consilient model of material processes, drawn from the pure and applied sciences, it is shown that information is not directly physically productive, but rather, is indirectly productive via second-law efficiency. In keeping with classical mechanics, energy and energy alone is physically productive. Information acts on material processes indirectly defining both the process and the outcome (good/service) and providing feedback on the former. The ICT revolution, by increasing transmission

speeds and lowering storage costs (two fundamental
ICT innovations), provided more and better quality
information which provided a potential for higher
second-law efficiency. However, because these gains
are one-shot in nature (given the overall stability
and upper limits of second-law efficiency), they
cannot be counted on to provide the sustained
increases in material wealth that were synonymous
with the first and second industrial revolutions.
The organization of work is affected; however, the
overall amount of work (material wealth) continues
to be governed by the laws of classical mechanics,
specifically those governing energy and work.

This is the basic finding of this book and the
rationale behind the title, namely that ICT has
ushered in, and continues to usher in a revolution
in the organization of work; however, it has not, is
not and cannot usher in an industrial revolution
comparable to those of the early 19th and early 20th
centuries. In short, energy and information are
incomparable in so far as classical mechanics go:
one is the basis of work and the cornerstone of all
material processes known to mankind, while the other
is definitional in nature.

The book is organized as follows. In Chapter 1,
we present our analytical framework, notably the
Energy-Organization (E-O) approach to modeling
material processes in general and in economics in
particular. The E-O approach has a definite
advantage over all others in that it explicitly
accounts for both energy and information, the main
ingredients of the first, second and possibly third
industrial revolutions. Moreover, it will allow us
to deconstruct the first and second industrial
revolutions into their component parts. Rather than
seeing both as a shift in an aggregate production
function, the E-O approach allows us to identify the
exact changes made in the relevant material
processes. This is extremely important to the
question at hand for the simple reason that before
we can compare general purpose technologies, we have
to know just what they are.

Chapter 2 uses the E-O approach to deconstruct
the first industrial revolution into three sub-
revolutions, each corresponding to a distinct steam-
based enabling technology. The first was the low-
pressure, condensing steam engine popularized by
James Watt and his business associate Matthew
Boulton. The second was the high-pressure, non-
condensing steam engine, while the third was the
steam turbine. With the introduction of each came a
new round of energy deepening and output growth, the
sum total of which defines the first industrial
revolution.

Chapter 3 does likewise in the case of the second industrial revolution. However, in this case, we identify two sub-revolutions, each being defined by a distinct enabling technology. The first-second industrial revolution starts at the beginning of the century and runs through to the post-WWII period. After WWII, power ratings of dynamos increased markedly, resulting in a golden period of energy deepening, one that witnessed unprecedented increases in conventionally-defined labor productivity and output growth. This golden age came to an abrupt halt in the mid-1970's with the two OPEC-induced energy crises, which, not surprisingly, brought to a halt all new, more energy intensive enabling technologies. The emphasis in material process engineering went from energy deepening to energy efficiency, two polar opposites. Chapter 4 examines the end of energy deepening and, as such, the de facto end of the second industrial revolution.

Having deconstructed both the first and second industrial revolutions in terms of (i) enabling technologies and (ii) energy deepening, we then turn in Chapter 5 to the ICT revolution, which we deconstruct into (i) an enabling technology in the form of computers-microchips and (ii) the associated information deepening. ICT increased agents and firms ability to process and store information. The problem is that information, unlike energy, is not physically productive. This then raises the question: what is the information revolution if not an industrial revolution? To shed light on this, we draw from the material process engineering literature as well as from a number of observers.

1

Analytical Framework

> The term process can in general be defined as a
> change in the properties of an object,
> including geometry, hardness, state,
> information content (form data), and so on. To
> produce any change in property, three essential
> agents must be available: (1) material, (2)
> energy, and (3) information.
> —Leo Alting, Manufacturing Engineering
> Processes

Introduction

While a valuable contribution to the growth
literature, work on General Purpose Technologies
(GPT's) suffers from one basic, overriding flaw,
namely it parameterizes technology and technological
shocks. The latter are not examined within a general
framework where the shock can be compared and
contrasted with other shocks, quantitatively and
qualitatively, and measured. Rather, they are
examined within standard neoclassical production
theory where technology is modeled parametrically,
specifically as a scalar. Reducing technology and
technological shocks to scalars precludes meaningful
comparisons.

This chapter, and indeed this book, takes issue
with this approach, arguing that while useful as a
first approximation, it is grossly insufficient and
inadequate. Factor inputs that are important in
process engineering and applied physics (energy and
information) are ignored, the focus being on capital
and labor.[1] Questions pertaining to energy and energy
use, or information and information use are ignored,
oftentimes for lack of an appropriate framework.

It is with this in mind that this chapter
presents an alternative approach to modeling
material processes, namely the energy-organization

(E-O) approach where the focus is on the two
"universal" factor inputs, namely broadly-defined
energy and organization. Broadly-defined energy
includes both animate and inanimate forms of energy.
By animate energy, it should be understood muscular
(human and animal) power; by inanimate energy, it
should be understood wind, fossil fuel-based,
hydraulic and nuclear power.   All production
processes involve the consumption of energy of
one type or another.[2] Organization will be defined
as the conception/design of, and the overseeing
(i.e. supervision) of energy-consuming (i.e.
entropic) production processes.   The development
of the steam engine by Papin, Savary, Newcommen and
Watt in the 17th and 18th centuries is an example of
the former, while its day-to-day operation is an
example of the latter.   In the natural world, the
design of and supervision of energy-consuming
processes is governed by forces which are not fully
understood.   Nonetheless, we shall refer to such
processes as naturally-occurring/spontaneous
entropic processes. What differentiates these from
man-made (i.e. anthropomorphic) entropic processes
is the form of organization, broadly defined. From
the Paleolithic era to the present, Homo
sapiens—neanderthalensis and sapien—have designed
and redesigned man-made entropic processes (i.e.
anthromorphic entropic processes).   For example, the
development of stone tools in the Paleolithic era
altered the very nature of work.   By reducing waste
(i.e. increasing efficiency), primitive tools such
as hammers and knives increased the amount of work
that could be accomplished for a given quantity of
muscular energy (brawn).   In fact, the development
and improvement of anthropomorphic entropic
processes is what defines various pre-historical and
historical eras (e.g. the stone age, the bronze age,
the machine age).[2]

The E-O Approach to Modeling Production Processes

The Physical and Economic Definitions of Work

The cornerstone of the E-O approach to modeling
production processes (Beaudreau 1998) is Sir Isaac
Newton's second law of motion, namely F=ma, where F
is force, m is mass, and a is acceleration.   Put
differently, a=F/m.   That is, acceleration is simply
force divided by mass.   The corresponding definition
of work is W=fd where W is work, f is force, and d is
distance: work equals force times distance.   Thus,
the greater is f, ceteris paribus, the greater is W.

The greater is d, ceteris paribus, the greater is W. In short, the more force exerted and the longer the distance (or time period) over which the force in question is exerted, the more work performed. Redefining work as output permits us to write the basic axiom of E-O production analysis, namely that for any given, well-defined man-made entropic process, output is an increasing function of energy consumption.

Contrast this with the standard, time-invariant definition of work found in political economy, namely that work is an increasing function of capital and labor (i.e. W=f(K,L)). Capital and labor, it is argued, produce output (value added). Both are assumed to be productive. Just what it is that capital and labor do in production processes is unspecified. Terms such as capital productivity and labor productivity, however, connote the idea that both somehow work. Inanimate forms of energy such as oil, gas and electricity consumption are assumed to be intermediate inputs, and, hence, are not productive in the conventional sense. Put differently, they are not factors of production. In short, capital and labor are assumed to add value to energy and other raw materials. Broadly-defined organization is also ignored. Production processes are assumed to exist. Management issues are, in general, ignored.

Clearly, the physical and economic definitions of work are worlds apart. For three centuries, physicists have focused on force and energy as the basis of work; economists, on the other hand, have focused on capital and labor. In physics, tools and machines (i.e. capital) modify and transmit force and energy, but are not, as such, a source of energy. In political economy, energy (inanimate energy forms) is viewed as an intermediate good, and, thus, is not productive in the traditional sense (value-adding).

A Physical Model of Production

Here, we begin with an in-depth look at the purely physical aspect of production, namely the relationship between work and energy. To this end, we start by defining production as the following functional relationship between work (i.e. output) and the force (i.e. energy) expended in the process.

$$W(t) = F[E(t)]; \quad F'[E(t)] > 0 \qquad (1.1)$$

where: W(t)=work in period t; and E(t)=energy (i.e. force) in period t.

Animate and Inanimate Forms of Energy

In general, energy can be disaggregated into two categories, namely $E_a(t)$, animate (i.e. muscular) energy, and $E_i(t)$, inanimate energy $[E(t)=E_a(t)+E_i(t)]$. Examples of animate energy include human and animal force (i.e. muscular force), while examples of inanimate energy includes internal combustion, steam power, wind power, hydraulic power and electrical power.[3] Work will as such be modeled as an increasing function of animate and inanimate force/energy.

The Role of Tools/Machines

From the Paleolithic era (i.e. the Stone Age) on, work as defined above, has invariably involved the use of tools/machines. These include axes, adzes, levers, presses, drills, screws, hammers, screwdrivers, saws, etcetera. This raises a number of questions. For example, what is the role of tools in broadly-defined work? Are they a source of energy? Are they productive in the physical sense? Similar questions were raised by Nobel Prize laureate (Chemistry) Frederick Soddy in the 1920's. Who or what, he asked, was ultimately responsible for producing goods and services? As the following quote indicates, in Soddy's mind, there was no doubt: energy was the primary factor.

> At the risk of being redundant, let me illustrate what we mean by the question "How do men live?" by asking what makes a railroad train go. In one sense or another, credit for the achievement may be claimed by the so-called "engine-driver," the guard, the signalman, the manager, the capitalist, the share-holder,-or, again, by the scientific pioneers who discovered the nature of fire, by the inventors who harnessed it, by labour which built the railroad and the train. The fact remains that all of them by their collective effort could not drive the train. The real engine-driver is the coal. So, in the present state of science, the answer to the question how men live, or how anything lives, or how inanimate nature lives, in the sense in which we speak of the life of a waterfall or of any other manifestation of continued liveliness, is, with few and unimportant exceptions, "By sunshine." Switch off the sun and a world would result lifeless, not only in the sense of animate life, but also in respect of by far the greater part of the life of inanimate nature. (Soddy 1924, 4)

Accordingly, because tools are not a source of energy, they are not productive in the physical sense. That is, they do not "work." Nowhere is this better seen than in the physicist's definition of a "machine" which consists of an instrument used to transmit or modify applied force/energy.

> Machinery is used to change the magnitude, direction and point of application of required forces in order to make tasks easier. The output of useful work from any machine, however, can never exceed the total input of work and energy. (Betts 1989, 172)

Arthur Beiser, in Modern Technical Physics, provides a similar definition:

> A machine is a device which transmits force or torque to accomplish a definite purpose. (Beiser 1983, 208)

By using primitive hammers and knives, early man (i.e. Paleolithic and Neolithic man) was better able to direct and apply his force. Analytically speaking, however, while tools allowed him to minimize energy loss (i.e. wasted energy), they did not allow him to increase the total amount of work beyond the initial level of force. That is, they were not a source of additional energy.[4] Implicit here is the basic notion of thermodynamic efficiency, defined as work out versus work in. By better transmitting muscular force, tools improved primitive man's thermodynamic efficiency. By expending the same amount of energy, more work could be done (e.g. skinning animals, cutting fire wood).
This raises the question of the relevant measure of tool/capital productivity. Specifically, how do we measure and define the productivity of tools/machines (i.e. capital)? The answer, we believe is relatively straightforward. Since tools/machines/capital are not a source of energy, they cannot be regarded as productive in the physical sense (i.e. performing work/working). Rather, capital productivity must be measured in terms of its ability to transform force/energy into useful work, a concept known as second-law efficiency. As such, the notion of capital productivity is a qualitative one (i.e. a scalar), and not a quantitative one. Increasing the amount of capital, ceteris paribus, will not increase output. However, increasing the quality of capital, say by improving thermodynamic efficiency, will increase output. For example, James Watt's steam engine, by reducing the heat loss common to the Savary and Newcommen condensing steam engines,

increased the overall efficiency (i.e. exergy) of
what at the time was known as "fire power." The
fact, however, remains that it was not, and is not,
an energy source. This allows us to rewrite Equation
1.1 as follows:

$$W(t) = \eta E(t) \qquad\qquad (1.2)$$

Where η is defined as second-law efficiency.[5]
Equation 1.2 defines the E-O approach to production
processes.   W(t), output/work is an increasing
function of E(t), energy/force, and η, thermodynamic
efficiency.   Increasing η, thermodynamic efficiency,
ceteris paribus, leads to higher output/work.
Increasing energy, ceteris paribus, leads to higher
work/output. η, it therefore follows, is tool
specific.   Better, more efficient tools will have
higher η's, and vice versa.

A Tool/Machine Taxonomy

According to classical mechanics, there are three
basic tools/machines: the lever, the inclined plane
and hydraulic press.[6]   Each directs and applies a
specific form of energy.   Table 1.1 lists the three
basic tools/machines and some of their uses.   We
see, for example, that the lever is the basic
tool/machine which transforms force in the case of a
crane, a wheel barrow, and various pulleys. Lastly,
there are what we choose to refer to as composite
tools/machines, consisting of combinations of the
three basic tools.   All tools/machines, it therefore
follows, should to be viewed as combinations of the
three basic tools, the lever, the inclined plane and
the hydraulic press.

Table 1.1
A Tool Taxonomy

Basic Tool	Composite Tool
Lever	Scissor, Plier, Wheelbarrow Pulley
Hydraulic Press	Hydraulic Jack
Inclined Plane	Wedge, Jack, Screw

An   Organization-Augmented   Physical Production   Model: The   E-O   Production Function

The discussion of the role of tools in modern production processes brings us to examine, in general terms, the second universal factor input, namely, organization. Machines define the framework in which force/energy is transformed into work. In physics as in economics, it is implicitly assumed that (i) well-defined work processes exist (i.e. Equation 1.2) and (ii) that they are self-regulating. That is, the very setting in which force or energy is transformed into work is well-defined (i.e. exists), and secondly, is non-stochastic. Machine breakdown, broadly defined to include such things as feedstock breakdown and mechanical failure, is ignored. In this section, we examine the broader question of organization. The discussion begins with a definition of what we shall refer to as anthropomorphic entropic processes. That is, man-made energy-using processes. This is then followed by two issues, namely (i) the design of and (ii) the supervision of man-made (i.e. anthropomorphic) energy-consuming processes. Together with tools, these define broadly-defined organization.

Natural and Anthropomorphic Entropic Processes

Our treatment of organization begins with a discussion of entropic processes in general. Entropic processes are defined as environments in which energy is transferred from one system to another or others in the form of work. For the sake of discussion, we define two basic types of entropic processes: natural entropic processes which occur spontaneously in nature, and anthropomorphic entropic processes which, as their name implies, are conceived of and overseen by man (i.e. managed, supervised).[7] For example, solar radiation, the wind, the tides, river currents, etcetera are examples of natural entropic processes based on solar energy. The production of food, tools, shelter, and culture, however, are examples of anthropomorphic entropic processes. What distinguishes these is the nature of the relevant set of instructions. In the former case, it has evolved randomly (or so it appears) over the course of the last 15 billion years, while in the latter, it has evolved consciously over the course of the last 2,000 years. In the case of anthropomorphic entropic processes, Homo sapiens-sapiens design and

supervise (i.e. oversees) the relevant work process.
Take, for example, the simple wind mill, which
converts the result of a temperature gradient (i.e.
the wind) into work. The human body, however, is
an example of a natural entropic process whose
blueprint is the work of millions of years of
evolution/mutation (i.e. the human genome), and
whose supervision is auto-regulated.

Clearly, tools are the defining element in
anthropomorphic entropic processes. They are to
production processes what the human body is the
process of life. They transmit and/or modify
force/energy. Given the presence of friction, it
follows that the better are the tools (i.e. the less
friction), the higher the efficiency, and the
greater the work.

The presence of energy and tools, however, is not
a sufficient condition for work to occur as both
occur spontaneously in nature. Also required is an
activity which we shall refer to as the "overseeing
of" or "supervising of" entropic processes. For
example, take the case of a simple gasoline engine-
powered water pump. With an endless supply of
refined hydrocarbons, it should work continuously ad
infinitum. Suppose, however, that, for some reason,
the pump arm spontaneously dislodges itself from the
drive shaft. Clearly, in this case, the energy
provided by the engine will be lost, unless of
course the problem is rectified. Supervision, as
this example illustrates, is an important aspect of
organization. Anthropomorphic entropic processes
are subject to breakdown.[8] Supervision, one could
argue, minimizes the resulting energy loss.
Nineteenth-century political economist Alfred
Marshall described the role of supervision in
"modern" manufacturing concerns in the following
terms:

> New machinery, when just invented, generally
> requires a great deal of care and attention.
> But the work of its attendant is always being
> sifted; that which is uniform and monotonous is
> gradually taken over by the machine, which thus
> becomes steadily more and more automatic and
> self-acting; till at last, there is nothing for
> the hand to do, but to supply the material at
> certain intervals and to take away the work
> when it is finished. There still remains the
> responsibility for seeing that the machinery is
> in good order and working smoothly; but even
> this task is often made light by the
> introduction of automatic movement, which
> brings the machine to a stop the instant
> something goes wrong. (Marshall 1890 , 218)

To capture the role of tools and supervision in production processes, Equation 1.2 is rewritten as follows:

$$W(t) = \eta \ [S(t),T(t)] \ E(t) \qquad (1.3)$$

where $S(t)$ and $T(t)$ correspond to the level of supervision at time t and tools at time t, respectively. For the time being, it will be assumed that second-law efficiency is an increasing function of tools and supervision. Thus, for a given quantum of energy, the more/better tools and supervision, the greater is second-law efficiency (i.e. η), and hence, the greater is output (i.e. $W(t)$). Capital and supervision, it therefore follows, are not directly productive; rather, their contribution to production is via second-law efficiency.

Here, we stop short of explicitly modeling supervisory activity, except to point out the obvious, namely that, historically, it has been carried out by people (animate supervision), and secondly, has been organized hierarchically, with conventionally-defined workers (lower-level supervisors) at the bottom, line supervisors above, and senior managers (e.g. CEO's, CFO's and the Board of Directors) at the top. It is clear that while not a source of energy, the supervisory input is nonetheless a sine qua non of virtually all production processes. Without supervisors, output becomes probabilistic (including the null set).[9] Tools and machines are apt to break down, resulting in a loss of energy and output.

Leibenstein's Managerial Taxonomy

This view of the organization input (broadly-defined management) as a two-dimensional activity (e.g. supervision and design) is not entirely new. In work on the role of entrepreneurs in economic development, Harvey Leibenstein identified two types of managers, namely routine and innovative. Put differently, some managers are concerned with the day-to-day functioning of the firm, while others are concerned with innovation whether it involves products or processes.

> I may distinguish two broad types of entrepreneurial activity: at one pole, there is routine entrepreneurship, which is really a type of management, and for the rest of the spectrum we have Schumpeterian or "new type" entrepreneurship. (We shall refer to the latter as N-entrepreneurship.) By routine

entrepreneurship we mean the activities
involved in coordination and carrying on a
well-established, going concern in which the
parts of the production function in use (and
likely alternatives to current use) are well
known and which operates in well-established
and clearly defined markets. By N-
entrepreneurs, we mean activities necessary to
create or carry on an enterprise where not all
markets are well established or clearly defined
and/or in which the relevant parts of the
production function are not completely known
(Leibenstein 1968, 72).

The former are the equivalents of the supervisors
referred to above, while the latter are the
equivalent of the "designers/conceivers," those who
conceive of or improve upon existing entropic
processes or products. Put differently, progress
requires dynamic innovative managers; otherwise, it
will be condemned to manage the existing set of
anthropomorphic entropic processes ad infinitum.
Recall that anthropomorphic entropic processes,
unlike natural entropic processes, are man-made,
subject to change if and only if changed.

The E-O Production Function

This simple two-factor model (energy and
organization), based on the laws of motion and
thermodynamics, and the principles of organization
(i.e. design and supervision) describes virtually
all material processes, past, present and future.
Take, for example, tree harvesting. Prior to the
advent of the mechanical saw, animate energy was
transmitted and applied by tools such as axes and
saws to fall trees. The required supervision was
typically provided by the owners of animate energy
(i.e. lumberjacks). With the advent of the chain
saw, animate energy (i.e. the internal combustion
engine) was combined with inanimate energy, with
essentially the same tools and supervision (i.e. saw
teeth), to accomplish the task. More recently,
with the advent of the mechanical harvester,
inanimate energy alone is used in conjunction with
tools to accomplish the task. The requisite
supervision is provided by the operator. As this
case clearly demonstrates, technological change has
resulted in a case in which labor goes from
providing energy and organization to providing only
organization.
Another example is weaving. In the Paleolithic
era, fibers or reeds were woven together to produce
clothing, baskets and sieves using animate energy
only. With the advent of the loom (i.e. tools),

human energy was better transmitted and applied, resulting in higher output. Primitive looms, it therefore follows, are examples of anthropomorphic entropic processes. The application of water, steam, and electric power to the hand- and foot-operated loom led to the power loom where human energy was replaced by water, steam or electric power, transforming labor's role from that of inanimate energy source and supervisor to that of supervisor with the result that today, little to no human energy is deployed in the weaving process. In the following passage, 19th century British economist Alfred Marshall describes just such a change:

> We may now pass to the effects which machinery has in relieving that excessive muscular strain which a few generations ago was the common lot of more than half the working men even in such a country as England . . . in other trades, machinery has lightened man's labours. The house carpenters, for instance, make things of the same kind as those used by our forefathers, with much less toil for themselves . . . . Nothing could be more narrow or monotonous than the occupation of a weaver of plain stuffs in the old time. But now, one woman will manage four or more looms, each of which does many times as much work in the course of a day as the old hand loom did; and her work is much less monotonous and calls for much more judgment than his did (Marshall 1890, 218).

The operative word here is "manage." Analytically speaking, factory workers in the 19th century no longer worked; rather, they managed.[10] Simon Newcomb, the 19th century physicist and part-time economist, described the far-reaching change in labor's role in production brought about by steam-powered machinery as follows:

> We may readily apply the principle here illustrated to the actual historical facts. The introduction of machinery during the last hundred years has to a certain extent changed the directions of man's occupations. Instead of making things with their own hands, as they formally had to do, they are now managing machines or assisting in various ways in working them. The pin-makers are no longer at work; a few of them are feeding pin-making machines, but the majority of them have learned other employments. A large class of carpenters no longer pushes the plane; a portion of them feed the planning machines, and the remainder is fully occupied in executing work that

increased   refinement   which   demand   has
encouraged.   The same thing may be   traced
all   through   the   channels   of industry
(Newcomb 1886, 390).

This was a recurrent theme in the 19th century,
extending beyond political economy into such fields
as engineering and mechanics.   For example, James
Martineau, a professional engineer, in a lecture to
the   Liverpool   Mechanics   Institute,   spoke   of
machinery "rapidly supplanting human labour and
rendering mere muscular force...worthless... That
natural machine, the human body, is depreciated in
the market.   But if the body have lost its value,
the   mind   must   get   into   business   without
delay"(Tylecote 1957, 262).
The depreciation of human "power" continued in
the early 20th century with the introduction of
electric power.   More flexible than steam power, it
resulted in the mechanization sectors of the economy
which had resisted earlier mechanization.   Consider,
for example, the following quote describing the
effects of electric-powered materials handling
equipment on the nature of work in the mining
industry, taken from David Nye's work on the
electrification of America (Nye 1990):

> Thus   for   some,   electrification   meant
> unemployment as a few skilled jobs replaced
> unskilled labor.   At Green Ridge Colliery, for
> example, a station engineer, motorman, and
> helper could run an electric locomotive that
> replaced six mule drivers, four boy helpers,
> and seventeen mules.   At New York and Scranton
> Coal, three men and a locomotive replaced seven
> boys and fourteen mules.   Down in the mines
> other electrical machines replaced hand labor,
> increasing productivity but reducing the need
> for artisanal skill.   Such innovations were
> contributory factors in the United Mine Workers
> strikes of 1900, 1902, and 1912. A similar kind
> of labor   replacement   disturbed   the   steel
> workers, as General Electric designed a set of
> electrical motors and controls for rolling
> mills.   The corporation proudly announced, "One
> man surrounded by a dozen or more operating
> levers controls every motion of steel from the
> ingot furnace to the completed rail.... Every
> motor replaced a man, but the work is done
> better and more quickly than formerly." In many
> mills, "a motor [was] installed on a charging
> and   drawing   machine,   arranged   so   as   to
> automatically grip and withdraw the hot bloom
> from the furnace, and release it when clear of
> the furnace door.   The motorman has simply to
> start and stop the motor." (Nye 1990, 206)

Substitutes or Complements?

The resulting two-factor (energy and organization) model of production processes provides a rich framework in which to study production-related questions. For example, it provides a theoretical basis for studying the role of conventionally-defined energy, capital, labor and organization in production. To what extent are energy and organization substitutable? Can managers be substituted for energy, and can labor be substituted for capital? Further, it provides the necessary framework to study technological change and its effects on the various factor inputs over time. For example, has second-law efficiency increased or decreased over time?

Another relevant question, especially in light of the energy crisis in the 1970's and 1980's, is whether tools (i.e. capital) and broadly-defined energy are substitutes or complements? Most political economists felt that they were substitutes (Solow 1974). E-O production analysis provides valuable insights into this unresolved issue. The answer is a qualified no. Theoretically, for a given value of η, an increase in capital (i.e. tools) cannot make up for a loss of energy. Referring to Equation 1.3, we see that the only way in which more capital could compensate for less energy is if the result was higher second-law efficiency. That is, only if the additional capital increases η will output increase. As pointed out earlier, because machines and tools are not a source of energy, but rather, apply and transmit energy, it follows that, ceteris paribus, quantities of capital and energy cannot possibly be substitutes, but rather, are complementary inputs. Tools and machines cannot create energy. More tools for a given level of energy will not result in more work.

This leaves the possibility of second-law efficiency-increasing investment. In this case, the level of second-law efficiency is increasing in broadly-defined capital, measured by the value of tools and machines. That is, the more capital, the higher is η. Were this to hold empirically, then capital and energy would be substitutes in the sense that more capital, by providing for higher η values, reduces the amount of energy required to perform a fixed, given amount of work. The point is that capital and energy can only be substitutes if and only if the former is heterogeneous in nature. More efficient capital can, at least theoretically, compensate for less energy; otherwise, they are complementary.

E-O Productivity Measures

The E-O Production Function provides important insights into the definition, the construction and the use of productivity measures. Typically, productivity is measured as the ratio of output (value added) to one or many inputs. For example, in the past, political economists have defined and measured labor and capital productivity. Labor productivity is the ratio of output (value added) to labor input; capital productivity is the ratio of output (value added) to capital input. As pointed out above, implicit in these notions is the fact that both capital and labor are productive. This raises an important question. Does the fact that an input is physically present necessarily imply that it is productive? Take, for example, the case of managers. Are managers productive? If so, then how should their productivity be measured? After all, managers are a sine qua non of production. The point is that productivity measures differ both with regard to their content and, of course, with regard to their meaning. Those that refer to energy and force are to be viewed as physically productive, while those which refer to organization are to be viewed as organizationally productive. The fact that labor productivity increases does not imply that labor is responsible for the increase. As argued above, the only physically-consistent measure of productivity is the ratio of output to total energy.

Energy-Related Productivity Measures

Three energy-related productivity measures follow. Referring to Table 1.2, the first is the ratio of output to animate energy (i.e. $W(t)/E_a(t)$), commonly known as labor productivity. For this ratio to be relevant (i.e. in terms of thermodynamics), it must be the case that labor is a source of energy. Otherwise, labor productivity as conventionally defined is devoid of any meaning. As we shall argue later, labor in modern production processes is more appropriately viewed as a form of lower-level organization (i.e. supervisor). The second productivity measure is the ratio of output to inanimate energy (i.e. $W(t)/E_i(t)$), $E_i(t)$ being broadly defined to include electrical, thermo-mechanical, and all other inanimate forms of energy. By virtue of Equation 1.1, this measure explicitly defines η, thermodynamic efficiency. Lastly, the third law of thermodynamics provides us with a third measure of energy-based productivity, namely the ratio of output to total

energy (i.e. $E(t)$). This measures the relationship
between output and the broadly defined energy input.
Clearly, energy-conversion tables (i.e. labor-BTU,
electric power-BTU) are required to render this
measure operational. However, for production
processes requiring no animate energy (i.e.
$\dot{E}_a(t)$), this ratio is equivalent to the output-
inanimate energy ratio.

Table 1.2
E-O Analysis-Based Energy Productivity Measures

$W(t)/E_a(t)$	Ratio of Work to Animate Energy
$W(t)/E_i(t)$	Ratio of Work to Inanimate Energy
	(i.e. η)
$W(t)/E(t)$	Ratio of Work to Total Energy

Organization-Related Productivity Measures

As pointed out above, continued energy-deepening
throughout the 19th and 20th century altered
fundamentally the supervisory input (i.e. $S(t)$). In
the pre-industrial revolution period, the
organizational input (i.e. $S(t)$) and the source of
energy (i.e. $E(t)$) were one, namely the worker.
Weavers, stonecutters, shoemakers, and artisans in
general provided both energy and organization. In
time, as inanimate energy replaced animate energy,
the organizational aspect of production changed. A
new class of production-related personnel came into
being, namely managers. Together with workers who
had been relieved of the physical demands of work,
managers supervised production processes. With this
was born the fully-functional organization
hierarchy, with workers (lower-level supervisors) at
the base, and senior managers (upper-level
supervisors) at the apex. To capture the
developments in supervision technology, we
define two sub-classes of supervisors: $S_l(t)$,
lower-level supervisors (i.e. workers) and $S_u(t)$,
upper-level supervisors, where $S(t)=S_l(t)+S_u(t)$. The
former find themselves at the base of the
organizational hierarchy, while the latter find
themselves at the apex.
Table 1.3 presents the resulting four
organization-related productivity measures, namely
the ratio of work to total supervisors, the ratio of
work to lower-level supervisors, the ratio of work
to upper-level supervisors, and lastly, the ratio of
work to tools and machines. The latter is commonly
referred to as capital productivity. Ideally, the
capital data used in constructing this productivity
measure should include only the relevant tools and

machines. That is, they should be net of all energy-related capital (e.g. steam boilers, transmissions, transformers, capacitators, etcetera) and net of what we refer to as supervisory capital (e.g. control devices, computers, etcetera). Including energy-related capital would result in double counting.

Table 1.3
E-O Analysis-Based Organization Productivity Measures

$W(t)/S(t)$	Ratio of Work to Supervisors
$W(t)/S_1(t)$	Ratio of Work to Lower-Level Supervisors
$W(t)/S_u(t)$	Ratio of Work to Upper-Level Supervisors
$W(t)/T(t)$	Ratio of Work to Tools and Machines

E-OPA and Empirical Work

Energy-Organization production analysis (E-OPA) has important implications for empirical work. From Charles Cobb and Paul Douglas's first attempt at estimating output elasticities and, hence, marginal revenue products, to the current practice of estimating flexible-form production functions, the underlying assumption has been that capital and labor are physically productive. Clearly, as argued here, this assumption is unsubstantiated. Paleolithic production processes notwithstanding, capital and labor are organizational variables. Capital and labor output elasticities are, as such, devoid of any meaning. Only energy is productive in the traditional sense.

Such a view was borne out by earlier empirical work on the KLEP (capital, labor, electric power) production function, where, using data from U.S., German and Japanese manufacturing, we found high output elasticities for electric power, and low output elasticities for capital and labor (Beaudreau 1995a,b,c,1998).

This raises the problem of estimating production functions. Specifically, how should energy and organization be treated? Should labor and capital, for example, be included along with energy? Should they be nested in estimates of η, second-law efficiency, the idea being that they affect productivity indirectly via η ? We shall return to this later.

Conclusions

Finding mainstream production theory to be deficient in its treatment of energy and information in modern material processes, this chapter has presented an alternative analytical framework, one based on the author's previous work on production and distribution, and, to a lesser degree, exchange. Production was modeled as an increasing function of second-law efficiency, itself a function of tools (capital) and supervision (labor), and energy consumption. The former were referred to as organization-related factor inputs, while the latter was referred to as an energy-related factor input. The result is an general model of production and growth, one that both motivates and guides the re-examination of ICT in production processes and as the purported basis of the third industrial revolution. Specifically, it allows us to deconstruct the first and second industrial revolutions and examine in detail the technological underpinnings. Further, it allows us to address the question at hand, namely can/will ICT prompt an industrial revolution?

2

The First Industrial
Revolution Deconstructed

> It is well known that, during the last half
> century in particular, Great Britain, beyond
> any other nation, has progressively increased
> its powers of production, by a rapid
> advancement in scientific improvements and
> arrangements, introduced, more or less, into
> all the departments of productive industry
> throughout the empire. The amount of this new
> productive power cannot, for want of proper
> data, be very accurately estimated; but your
> Reporter has ascertained from facts which none
> will dispute, that its increase has been
> enormous—that, compared with the manual labour
> of the whole population of Great Britain and
> Ireland, it is, at least, as forty to one, and
> may be easily made as 100 to one; and that this
> increase may be extended to other countries;
> that it is already sufficient to saturate the
> world with wealth and that the power of
> creating wealth may be made to advance
> perpetually in an accelerating ratio.
> —Robert Owen, Report to the County of
> Lanark

Introduction

The shift from the domestic system of production to
the factory system, based on new sources of power,
had a profound impact on the industrial and economic
landscape of the United Kingdom. For example, the
ownership of factor inputs (energy and organization)
was no longer concentrated (i.e. within the family
unit), but diffuse. Workers that had up until then
been a source of motive power were reduced to
supervisory inputs (i.e. machine operatives),
overseeing the workings of steam-powered machinery.
Producers no longer coordinated production activity
in house, but rather is so in a market environment.
So great were these changes, we maintain, that moral
philosophers and political economists would devote

the next hundred years to understanding them, with,
as shall be shown, mixed results.

In this chapter, we reexamine the first
industrial revolution through the prism of the E-O
approach to material processes. More specifically,
the first industrial revolution will be
deconstructed into three sub-revolutions, each
defined by a steam technology. The first is the
standard, Watts-Boulton, external condenser, steam
engine, the flagship of the industrial revolution.
The second is the high-pressure steam engine
developed in the 1840's, making for a series of
speed-ups. The last is the development of steam
turbine by Charles Parsons in the 1860's which
reduced the associated capital costs and increased,
yet again, machine speeds and productivity.

Having deconstructed the first industrial
revolution into three sub-revolutions, we then turn
and examine the associated history of thought,
specifically, the history of the steam engine as
seen by early political economists (moral
philosophers). It will be shown that problems
associated with new technology and not the
technology itself are what prompted an interest.
Consequently, given that the focus was on the
problems, any deeper understanding of the technology
was precluded. This, we argue, ultimately
contributed to the current underdeveloped state of
knowledge of the steam engine as a GPT.

Steam Power and the Industrial Revolution

This section is devoted to evaluating the effects
of "fire power" on the creation of material wealth.
Particular attention will be paid to the various
problems "fire power" raised. Among these are the
problems of exchange and distribution.

The Domestic System

To better understand and appreciate the impact of
"fire power" on industry in the United Kingdom, we
begin by describing production, distribution and
exchange activity in the years leading up to the
industrial revolution. Prior to the introduction of
the Watt-Boulton steam engine, work (i.e. output) in
Great Britain was, for the most part, the result of
the application of muscular (human and animal)
force. Draft animals, supervised by farmers,
provided the necessary force to till the soil.
Human, muscular force, however, was required to
harvest crops. Clearly, the stronger the draft
animals, the more work could be done per unit of
animal feed. The heightened interest in animal

breeding in the pre-industrial period can, as such, be viewed as a means of increasing the relevant η, the level of thermodynamic efficiency. The same is true of the tool industry, where the emphasis was on high-grade, durable steels, capable of doing more work per unit force—in this case, muscular.

In 17th century Great Britain, most forms of production activity were carried out on a small scale, often times by the household unit. Food was grown, wool was produced, thread was spun, cloth was woven, and tools/furniture were fashioned. Among the specialized activities were steel making, and, to a lesser degree, milling and, of course, religion—salvation. Autarky was, in general, the rule. Exceptions, however, involved the production of goods, most often textiles, by households for the market. Merchants from cities would contract work out to households in the surrounding countryside, supplying the raw material and purchasing the finished products. This is referred to as the "domestic system," "putting out system," and the "cottage industry." Peter Stearns describes this period as follows:

> With all these developments, however, Western technology and production methods remained firmly anchored in the basic traditions of agricultural societies, particularly in terms of reliance on human and animal power. Agriculture had scarcely changed in method since the fourteenth century. Manufacturing, despite some important new techniques, continued to entail combining skill with hand tools and was carried out in very small shops. The most important Western response to new manufacturing opportunities involved a great expansion of rural (domestic) production, particularly in textiles but also in small metal goods. Domestic manufacturing workers used simple equipment, which they usually bought themselves, and relied on labor from the household. Many combined their efforts with farming, and in general their skill levels were modest. The system worked well because it required little capital; rural households invested a bit in a spinning wheel or a hand loom, while an urban-based capitalist purchased the necessary raw materials and, usually, arranged for the sale of the product. Output expanded because of the sheer growth of worker numbers, not because of technical advancement; indeed, the low wages paid generated little incentive for technical change. (Stearns 1993, 18)

What is particularly interesting, as far as we are concerned, is the nature of production, exchange

and distribution. The household unit (producer)
provided the energy, tools and supervision necessary
to transform say raw wool into thread, thread into
finished cloth, or, for that matter, finished cloth
into clothing. The energy was, in general, muscular
in nature; the tools were primitive by today's
standard, and the supervision was provided by the
worker.[1] Given that the energy and organization in
question were owned by the household (i.e. the
household unit), there was no distribution problem--
-as described earlier (i.e. outside of the
household). The income received from the merchants
(payments for services rendered) went to the
household (producer), not to the individual factors,
per se.[2]

     The resulting exchange technology did not pose
any problems. The overall demand for trade credit
was, but for the merchant's profit, equal to the
value of total output. Trade credit was provided by
local and regional banks. The relevant credit
instruments included bills of exchange, promissory
notes, et cetera. Merchants demanded trade credit
in the amount of the cost of their raw materials
(cotton, wool, linen), and the cost of
transformation (adding value).

     Not surprisingly, the domestic system went
largely unnoticed, not being the subject of inquiry
of any natural or moral philosopher in the pre-
industrial period. The supply of human, animate,
muscular energy put an upper limit on the per-capita
amount of output/work, and, hence, of the growth of
output and work. One could argue that in the
absence of the steam engine, growth would have been
limited to growth of animate, muscular energy,
which, at the time, tracked population growth.

Enter the Steam Engine

     Two events in 18th-century Great Britain changed
the industrial landscape forever. The first was
Richard Arkwright's water-wheel-driven spinning
jenny, while the second was the development and
widespread application of the Watts-Boulton rotary-
power steam engine. Analytically-speaking,
Arkwright's water-powered spinning jenny was of
monumental importance, as it altered fundamentally
the nature of work. The force required to spin
cotton and wool would, from this point on, no longer
be animate, but inanimate. Spinning wheels
(spindles) would be powered by hydraulic power,
transmitted by way of complex belting, gearing and
shafting. This de facto lifted the energy
constraint. Never again would human muscular power

limit the amount of work done.   According to Peter
Stearns:

> More impressive developments occurred in the
> preparatory phases in cotton, James Hargreaves
> invented a spinning jenny device about 1764,
> which mechanically drew out and twisted fibers
> into threads—though this advance too initially
> was applied to handwork, not a new power
> source. Carding and combing machines, to ready
> the fiber prior to spinning, were developed at
> about the same time.   Then in 1769, Richard
> Arkwright developed the first water-powered
> spinning machine; it twisted and wound threads
> by means of flyers and bobbins operating
> continuously.   These first machines were
> relevant only for the cheapest kind of thread,
> but other inventions by 1780 began to make
> possible the spinning of finer cotton yarns.
> These new devices also could be powered by
> steam engines as well as waterwheels. The
> basic principles of mechanized thread
> production have not changed to this day, though
> machines were to grow progressively larger, and
> a given worker could tend to a number of
> spindles. (Stearns 1993, 23)

However, by moving production out of cottages
(putting-out) into factories (gathering-in), it
raised the problem of exchange and distribution.
How would total value added be apportioned among the
owners of capital (i.e. the mill), the owners of
lower-level supervisory skills (i.e. men, women and
children), and the owners of the hydraulic resource?
Moreover, given that the producer (manager) was, in
most cases, also the owner of the capital (equity
capital), a non-negligible fraction of factor
payments would be residual in nature, paid out after
the finished output was sold to merchants.

It is important to point out that these epoch-
defining changes predate the introduction of the
steam engine in industry. A good example of this is
Matthew Boulton's Soho Manufactory in Bolton where
the Hockley Brook powered the machines used to
manufacture buttons, watch chains, plated wares,
etcetera. Earlier in 1765, he had left Birmingham to
avoid operating an expensive horse mill.

Hockley Brook, however, was not without problems
of its own. For example, in the summer, water
levels were, often times, insufficient to power the
wheels, bringing the manufactory to a halt.   To
overcome this problem, Boulton turned to steam
power, ordering a Savery engine to pump water and,
in the process, turn the water wheels. According to
Richard Hills:

> In light of subsequent history, one of the most
> important people to consider using a Savery
> engine was Matthew Boulton when he wished to
> supplement the water resources of the Hockley
> Brook which powered his new Soho Manufactory.
> He had purposely moved out of Birmingham in
> 1765 to avoid operating an expensive horse
> mill, but the growth of the Manufactory
> presented him, with a similar dilemma once
> more. So his mind turned to employing a steam
> engine to lift water from the tail race of the
> waterwheel back to the mill dam. (Hills 1989,
> 40)

Realizing that steam could drive machinery
directly—as opposed to via waterwheels—Boulton
began work with James Watt in 1768 on what was to
become the defining energy innovation of the 18th
century, namely the Boulton-Watt rotary-drive steam
engine. These developments, we argue, are essential
to understanding Adam Smith's An Inquiry into the
Nature and Causes of the Wealth of Nations.
According to the record, Smith knew Boulton and
Watt, and had personally visited the Soho
Manufactory prior to 1776.

The late 18th century/early 19th century, it
therefore follows, was a transitional period, with
animate, muscular energy being replaced by
inanimate, hydraulic-based energy, which, in turn,
was replaced by inanimate, fossil-fuel-based energy
in the form of steam power. According to Katrina
Honeyman:

> During the period 1780-1825, cotton factories
> varied not only in scale but also in type. S.
> D. Chapman has identified what he sees as three
> main types. The Type A mill was a small-scale
> operation, often employing hand-operated
> jennies or mules and possibly horse capstans
> for driving card machines. The cost of
> establishing and equipping this type of mill
> was about £1,000 £2,000. The Type B factory
> was more often purpose-built and comprised
> three or four storeys. It came in two sizes:
> one designed to hold approximately 1,000
> spindles and requiring up to £3,000 investment;
> and one at least twice as large with up to
> 3,000 spindles, costing from £5,000. These are
> often referred to as Arkwright-type mills.
> Type C was larger, generally steam-powered, and
> cost about £10,000. This category of mill was
> not usual until the early 19th century.
> (Honeyman 1982, 57)

The substitution of inanimate power for animate,
muscular power occurred over a period of roughly 100
years (1780 1860). According to data on workers in

the U.K from 1806 to 1862, there were as many
factory workers as handweavers, and that by 1862,
the latter had, for all intents and purposes,
disappeared from the industry. Inanimate energy had
completely   displaced   animate   energy.
Industrialization being born of inanimate energy—in
this case, the consumption of coal—it follows that
the rate of growth of output would be an increasing
function of the rate of growth of coal consumption,
ceteris paribus.   Table 2.1 below reports both the
level of and rate of growth of coal production in
the United Kingdom at five-year intervals from 1755
to 1900.   From a level of 4.230 million tons in
1755, it had increased 12,049 percent to 50.968
million tons in 1850.   By 1870, it had more than
doubled to 110.4 million tons. At the end of the
century, it had more than doubled again to 225.[2]
million tons.

Table 2.1
Total Coal Consumption, Great Britain 1755–1900

Year	Coal Cons	Growth Rate*
1760	4,520	
1765	4,950	0.095
1770	5,520	0.115
1775	6,120	0.108
1780	6,750	0.102
1785	7,550	0.116
1790	8,570	0.145
1795	9,570	0.116
1800	10,960	0.145
1805	12,960	0.182
1810	14,790	0.141
1815	16,590	0.121
1820	18,900	0.139
1825	20,900	0.105
1830	24,800	0.186
1835	29,560	0.191
1840	35,270	0.193
1845	41,706	0.182
1850	50,968	0.222
1855	64,500	0.265
1860	80,000	0.240
1865	98,200	0.227
1870	110,400	0.124
1875	133,300	0.207
1880	147,000	0.102
1885	159,400	0.084
1890	181,600	0.139
1895	189,700	0.044
1900	225,200	0.187

*5-Year Growth Rate
Source: Mitchell (1988).

Referring to Column 3, we see that five-year growth rates ranged from a low of 6.8 percent in 1755 1760, to a high of 24.0 percent in 1855 1860. Given their broad nature, these data must be interpreted with caution. Ideally, data on industrial consumption of coal in the United Kingdom, not the overall total output of coal, would be required. Such data, unfortunately, do not exist. What do exist, however, are data collected by the Factory Inspectors, in accordance with the Factory Act of 1833 on the use of steam and water power in various U.K. industries from 1835 onward. These data, in combination with employment data also contained in, provide useful information on the extent of inanimate energy deepening in U.K. industry, especially in the textiles sector.

Table 2.2
Total Power Consumption, Textiles, Great Britain
1835–1903

Cotton

Year	Steam	Water	Total	$S_c(t)$	$E_c(t)/S_c(t)$
1838	46	12	58	259	0.2239
1850	71	11	82	331	0.2477
1856	87	9	96	379	0.2532
1861	281	12	293	452	0.6482
1867	190	12	202	401	0.5037
1870	299	8	307	450	0.6822

Wool

Year	Steam	Water	Total	$S_a(t)$	$E_w(t)/S_w(t)$
1838	17	10	27	87	0.3103
1850	23	10	33	154	0.2142
1856	31	9	40	167	0.2395
1861	53	11	60	173	0.3699
1867	85	12	97	262	0.3702
1870	103	12	115	239	0.4811

Miscellaneous

Year	Steam	Water	Total	$S_m(t)$	$E_m(t)/S_m(t)$
1838	7	4	11	42	0.2619
1850	11	3	14	68	0.2058
1856	14	4	18	80	0.2250
1861	32	4	36	94	0.3829
1867	42	5	47	135	0.3489
1870	52	5	57	146	0.3904

Source: Mitchell (1988).

Table 2.2 presents data on power employed, measured in horsepower, in the cotton, wool, and flax/jute/hemp/Chinagrass factories from 1835 to 1870, as well as the corresponding employment data. The sources of both series are the various returns filed by the Factory Inspectors and published in the Sessional Papers of the House of Lords. Before

turning to these data, it bears reminding that
workers in factories no longer exerted themselves
physically as was the case in the domestic system,
but, instead, supervised the spinning machines and
looms (power looms). Some, however, were assigned
the task(s) of managing feedstocks (inputs and
outputs). As such, by dividing total horsepower say
in the cotton industry by the total employment, one
arrives at a crude measure of inanimate energy per
lower-level supervisor. That is, the amount of
energy—and hence, work—workers supervised. The
evolution of this ratio over time provides a measure
of energy deepening.

Take, for example, the cotton industry, which in
1838, had 46,000 horsepower of steam power, and
12,000 horsepower of water power, for a total of
58,000 horsepower of inanimate energy. Given the
presence of 259,000 workers, it stands to reason
that the relevant $E_c(t)/S_c(t)$ ratio stood at
0.223938. This is to say that each worker
supervised the equivalent of 0.223938 of a
horsepower of inanimate energy. Referring to Column
4, we see that this ratio increased monotonically in
time, from 0.223938 in 1838 to 0.682222 in 1870.[3] In
practical terms, this implies that supervisors
(workers) in the cotton industry were overseeing
three times as more inanimate power in 1870 than in
1838, a span of only 32 years. As power is
synonymous with work, it follows that they were
supervising three times more work in 1870 than in
1838.

This jump in inanimate power consumption per
worker can be attributed to two factors, the first
of which being the declining use of handlooms,
referred to earlier. Animate energy as a power
source declined steadily from 1838 and, by 1870, had
virtually disappeared. The second factor is the
various technological improvements in the generation
of steam power. Foremost among these is the
development by Richard Trevithick and Arthur Woolf
of the high-pressure, non-condensing steam engine.
The latter increased the "indicated" horsepower of
existing and new engines far beyond their rated
"nominal" horesepower.[4]

Despite the dearth of data on output growth in
various British industries for the period under
consideration (1800–1900), what is clear is that
work as defined in the E-O sense increased at
unprecedented rates in the United Kingdom. Never
before had a country, nation, or people increased,
in so little time, the amount of work performed.
All of this, of course, was achieved by tapping into
a store of energy 450 million years old, in the form
of coal (Carboniferous Era). Not only did the
consumption of coal increase in this period, but the

efficiency with which it was used also increased,
increasing manifold the ability of the United
Kingdom to do work.

The Importance of Speed

The manifold increase in U.K. manufacturing
output in the 19th century resulted from, as argued
here, a process of energy deepening which
characterized much of this period. The energy
deepening, in question, was achieved in two distinct
ways, namely intensively and extensively. Starting
with the latter, extensive energy deepening refers
to the increase in the number and capacity of
spindles and looms in the cotton industry. With
Arkwright's spinning jenny, the number of spindles
in the U.K. cotton spinning industry increased
manifold. Similarly, with the advent of power
looms, the weaving capacity, as measured by the size
and number of looms, increased manifold.

A second source of energy deepening was the many
speed-ups that characterized the U.K. textiles
industry in the 19th century. Existing machines, be
they spinning or weaving, were speeded up, making
for greater output per period of time.
Characteristic of such "speed-ups" is an increase in
power consumption per period in time, in keeping
with the basic premises of E-O production analysis.
Improvements in steam engines throughout the 19th
century, along with improvements in spinning and
weaving per se, paved the way for increases in the
throughput rates per unit of capital (i.e. spindle
or loom). Among these was the development and
commercial use of the high-pressure steam engine by
Richard Trevithick and Arthur Woolf, which increased
markedly the speed of execution, and consequently,
the rate of output (throughput).

References to speed-ups in the early 19th century
are few and far-between. However, the few that do
exist convey the essence of the power revolution
that was playing itself out in Great Britain. Take,
for example, William Longston's testimony on working
conditions in the textiles industry before the
Committee on the Factories Bill.

> 9397. Is the intensity of application and of
> labour altered, either against or in favour of
> the operative and of the children employed in
> mills and factories?—I was a great number of
> years out of any factory, but those who were my
> acquaintances during my boyhood have often
> conversed with me, and they very frequently say
> that it cannot be less than double in intensity
> and exertion of physical application.
> 9398. State why you believe that the labour of

those employed has doubled since the first
introduction and use of cotton machinery, or at
least since you first knew it? The reason why
we believe so is from some calculations which
we have been obliged to make, and by my own
observation during the time we was manager of a
mill in 1830 and 1831, when we had some of the
same operations under my own observation.
9399. Have you any objection to put in those
calculations? Certainly not.

[The following document was then put in and
read.]

9400. It appears by this document that the work
done is very greatly increased between the
years 1810 and 1832; has the machinery been so
altered as to produce that amazing difference,
or does it result from accelerating the speed
of the machinery? It is from accelerating the
speed generally; and another cause is, that
more and more exertion is required from the
individual working at the machine; these are
the two causes.
9401. Those two causes, then, prove what you
have been asserting, namely that double the
labour and attendance is now requisite that was
formerly required? We think so.
9402. In spite of the improvements of
machinery? Yes; we believe those that are now
working in the same employment as we did when
we was a boy, do double the work.
9403. So that there may be a great improvement
in machinery, and at the same time a great
increase in actual labour to each operative?
Yes.
9404. Has that been, as far as your experience
has extended, the consonant result of
improvements of machinery in those mills,
namely, that the labour of the hands has
increased with every improvement in the
machinery; rather than diminished? Yes; the
improvements in the machinery have been great,
and the same physical exertion, and the same
physical exertion, and the same attention as
was formerly applied, would certainly produce,
in proportion to the altered state of
machinery, a much greater quantity and better
articles; but, added to that, the increased
exertions make the quantities to be such as
just now surprised you. (Committee on
Factories Bill 1832, 430)

As is clear from his testimony, machinery speed-
ups increased considerably the amount of "work"
demanded of the corresponding "lower-level"
supervisors. This was the direct result of the fact
that not all processes were automated. Faster
turning spindles and faster operating looms required
more exertion on the part of the "operatives"—that

is, those who supervised the machinery (machine
operatives). As the data presented show, output per
lower-level supervisor (stretcher) increased by 200
percent from 1810 to 1832. In the case of mule
yarn-spinning, the number of hanks spun, for 480
spindles ceteris paribus, increased by 50 percent
between 1806 and 1832.

The Role of Capital and Labor

The shift from the domestic system to the factory
system altered the nature of both capital and labor.
Labor, a source of energy (animate, muscular) and
thus of motive drive in the domestic system, was
reduced to a supervisory input, overseeing the
workings of continuous-flow machinery. The men,
women and children who, in the domestic system, had
spun and woven cotton, wool, linen, etcetera, would,
from now on, supervise the workings of steam-power
driven spinning and weaving machines, and managing
the various feedstocks (inputs and outputs) in
factories. Alfred Marshall would describe this far-
reaching change in 1890 in the following terms:

> We may now pass to the effects which machinery
> has in relieving that excessive muscular
> strain which a few generations ago was the
> common lot of more than half the working men
> even in such a country as England . .
> . in other trades, machinery has
> lightened man's labours. The house
> carpenters, for instance, make things of
> the same kind as those used by our
> forefathers, with much less toil for
> themselves . . . . Nothing could be more narrow
> or monotonous than the occupation of a weaver
> of plain stuffs in the old time. But now,
> one woman will manage four or more looms,
> each of which does many times as much work
> in the course of a day as the old hand loom
> did; and her work is much less monotonous
> and calls for much more judgment than
> his did (Marshall 1890, 218).

The key word here is "manage." The woman he
refers to now "manages," and, hence, no longer
weaves. Moreover, as he points out, each power loom
"does many times as much work in the course of a day
as the old hand loom did." Conventionally-defined
labor productivity, it follows, increased manifold.[5]
E. Baines, in the History of the Cotton Manufacture
published in 1835, refers to a tenfold increase in
productivity for weavers.

> A very good hand weaver, 25 or 30 years of age,
> will weave two pieces of 9-8ths shirting per

week, each 24 yards long, containing 100 shoots
of weft in an inch; the reed of the cloth being
a 44 Bolton count, and the warp and weft hanks
to 40 hanks to the lb.
In 1823, a steam-loom weaver, about 15 years of
age, attending two looms, could weave seven
similar pieces in a week.
In 1826, a steam-loom weaver, about 15 years of
age, attending to two looms, could weave twelve
similar pieces in a week; some could weave
fifteen pieces.
In 1833, a steam-loom weaver, from 15-20 years
of age, assisted by a girl about 12 years of
age, attending to four looms, can weave
eighteen similar pieces in a week; some can
weave twenty pieces. (Baines 1835, 240)

However, at this point in time, an important
distinction is in order, namely, that between
energy-based measures of productivity and
organization-based measures of productivity. In the
domestic system, workers physically exerted
themselves, spinning and/or weaving, in addition to
supervising the workings of their animate energy
(i.e. their body) and the tools (rudimentary
spinning wheels and hand looms). In the factory
system, their job task (job definition) was reduced
considerably. Specifically, it would consist almost
uniquely of supervision. As such, it follows that
the relevant productivity measure as found in the
factory system is organization-based (see Table 1.3
on page 20).
Capital also underwent a major transformation.
Specifically, the simple tools of the domestic
system (spinning wheels and hand looms) were fitted
with a power source (rotary power). Machinery was,
as a result, an amalgam consisting of rudimentary
tools fitted with an inanimate power source.
Complex shafting, belting and gearing systems were
used to transmit the torque generated by steam
engines to the spindles and the looms. Capital
costs per spindle or per square yard of weaving
capacity increased substantially, owing to power-
related costs, and to the need for more reliable
spindles and looms.[6] Under the domestic system,
spinning wheels and hand looms were fashioned out of
wood (hardwood and softwood), harvested, in most
cases, in local forests. The demands of high-
throughput, continuous-flow production, however,
were such that metal spindles and looms were needed,
increasing, as such, the cost of capital. This, as
it turns out, was true of all steam power-driven
industries.
As we shall argue later, these costs figured
prominently in the works of classical political
economists. The new drive technology required
massive investments, both in steam engines, and in

the accompanying tools. Saving and investment would,
over the next half century, occupy center stage in
political economy.

The Second-First Industrial Revolution

The second-first industrial revolution
corresponds to the period in the 19th century that
saw the development and implementation of a new
energy-consumption enabling technology in the form
of the high pressure steam engine. It is important
to remember that James Watt's "steam engine"
operated at low pressures and was of the condensing
variety. A number of developments in England and
America, specifically those by Trethivick and Evans,
resulted in the high-pressure (above 40
lbs/sq.inch), non-condensing engines. These so-
called high-pressure steam engines were more
efficient and more productive. Most importantly,
they paved the way for successive rounds of energy
deepening and productivity growth. We begin this
section with an excerpt taken from Wikipedia on
high-pressure steam engines.

High pressure engines

Around the year 1800, "high pressure" amounted
to what today would be considered very low
pressure, i.e. 40-50 psi (276 - 345 kPa), the
point being that the high pressure engine was
driven solely by the pressure of the steam and
once that steam had performed work, it was
usually exhausted at higher-than-atmospheric
pressure. The blast of the exhausting steam
into the chimney could be exploited to create
induced draught through the fire grate and thus
increase the rate of burning, hence creating
more heat in a smaller furnace.
The outcome was that engines could be made
much smaller than previously for a given power
output. There was thus the potential for steam
engines to be developed that were small and
powerful enough to propel themselves and other
objects. As a result, steam power for
transportation now became a practicality in the
form of ships and land vehicles, which
revolutionised cargo businesses, travel,
military strategy, and essentially every aspect
of society.
The importance of raising steam under
pressure (from a thermodynamic standpoint) is
that it attains a higher temperature. Thus, any
engine using such steam operates at a higher
temperature and pressure differential than is
possible with a low pressure vacuum engine. The
high pressure engine thus became the basis for
most further developments of reciprocating

steam technology.

Double-acting engines

The next major advance in high pressure steam engines was to make them 'double-acting'. In the single-acting high pressure engine above, the cylinder is vertical and the piston returns to the start (or bottom) of the stroke by the momentum of the flywheel.

In the double-acting engine, steam is admitted alternately to each side of the piston while the other is exhausting. This requires inlet and exhaust ports at either end of the cylinder (see the animated expansion engine below) with steam flow being controlled by valves. This system increases the speed and smoothness of the reciprocation and allows the cylinder to be mounted horizontally or at an angle.

Power is transmitted from the piston by a sliding rod sealed to the cylinder to prevent the escape of steam which in turn drives a connecting rod via a sliding crosshead). This in combination with the connecting rod converts the reciprocating motion to rotary motion. The inlet and exhaust valves have their reciprocating motion derived from the rotary motion by way of an additional crank mounted eccentrically (i.e., off-centre) from the drive shaft. The valve gear may include a reversing mechanism to allow reversal of the rotary motion.

A double-acting piston engine provides as much power as a more expensive 2-piston single-acting engine, and also allows the use of a much smaller flywheel than what would be required by a one-piston single-acting engine. Both of these considerations made the double-acting piston engine smaller and less expensive for a given power range.

Most reciprocating steam engines now use this technology, notable examples including steam locomotives and marine engines. When a pair (or more) of double acting cylinders, for instance in a steam locomotive, are connected to a common driveshaft their crank phasing is offset by an angle of 90°. This is called *quartering* and ensures that the engine will always start, no matter what position the crank is in.

Some marine engines have used only a single double-acting cylinder, driving paddle wheels on each side. When shutting down such an engine it was important that the piston be away from either extreme range of its travel so that it could be readily restarted (as there was not a second quartered piston to facilitate this).

The second-first industrial revolution witnessed significant energy deepening as firms replaced their Watt-Boulton engines with high-pressure ones. Referring to Tables 2.1 and 2.2 above, we see that energy consumption per worker ($E(t)/S(t)$) increased markedly in the 1860's, a period that witnessed an equally remarkable increase in overall coal consumption. Together, these increases define the resulting energy deepening. High pressure steam engines (enabling technology) increased the amount of energy consumed per worker and, in so doing, contributed to a marked increase in conventionally-defined labor productivity and consequently output.

The Third-First Industrial Revolution

As shown, the second-first industrial revolution was prompted by an innovation in the relevant enabling technology, one that allowed for higher levels of energy consumption, greater conventionally-defined labor productivity and greater output in general. In this section, we describe the third-first industrial revolution, namely the introduction of a yet another, more efficient energy enabling technology in the form of the steam turbine. We begin with a description of the steam turbine:

> A steam turbine is a mechanical device that extracts thermal energy from pressurized steam, and converts it into useful mechanical work. It has almost completely replaced the reciprocating piston steam engine, primarily because of its greater thermal efficiency and higher power-to-weight ratio. Also, because the turbine generates rotary motion, rather than requiring a linkage mechanism to convert reciprocating to rotary motion, it is particularly suited for use driving an electrical generator about 86% of the world's electricity is generated using steam turbine. The steam turbine is a form of heat engine that derives much of its improvement in thermodynamic efficiency from the use of multiple stages in the expansion of the steam, rather than a single stage.
> The first device that may be classified as a steam turbine was little more than a toy, the classic Aeolipile, created in the 1st century by Hero of Alexandria in Roman Egypt.[1][2][3] Heron's steam engine was also used to open temple doors and so was the first mechanical use of steam power. A thousand years later, a steam turbine with practical applications was invented in 1551 by Taqi al-Din in Ottoman Egypt, who described it as a prime mover for rotating a spit. Yet another steam turbine

device was created by Italian Giovanni Branca
in 1629. These early devices, however, were
very different from the modern steam turbine,
invented in 1884 by English engineer, Charles
A. Parsons, whose first model was connected to
a dynamo that generated 7.5 kW of electricity.
His patent was licensed and the turbine was
scaled up shortly after by an American, George
Westinghouse. The Parsons turbine turned out to
be relatively easy to scale up. Within Parsons'
lifetime the generating capacity of a unit was
increased by a factor of about 10,000.

A number of other variations of turbines
were developed that worked effectively with
steam. The de Laval turbine (invented by Gustaf
de Laval) places convergent-divergent nozzles
in between stages in order to extract more
energy from the steam.

The modern steam turbine has almost
completely replaced the reciprocating piston
steam engine (invented by Thomas Newcomen and
greatly improved by James Watt), primarily
because of its greater thermal efficiency and
higher power-to-weight ratio. In addition, the
turbine has only one moving part compared to a
piston engine, which can have dozens or even
hundreds. (en.wikipedia.org/wiki/Steam_turbine)

The steam turbine, like the high-pressure steam
engine, increased thermodynamic efficiency, raising
the amount of useful work that could be obtained
from a unit of energy and raising the amount of
energy consumed per period of time (unit of time),
both of which contributed to energy deepening. Set
in Equation 2.1, this amounts to an increase in E(t)
and consequently an increase in overall work
(material wealth).

It therefore follows that conceptually speaking,
the steam turbine is one of a number of energy-
related enabling technologies the sum of which
underlie the first industrial revolution. As such,
it was not the enabling technologies per se that
contributed to the manifold increase in productivity
and output, but the energy deepening they made
possible. The important point, in so far as this
book is concerned, is the fact that the first
industrial revolution was not a one-shot occurrence
but rather was a long, somewhat drawn out process
that spanned a century.

The First Industrial Revolution As Seen by 19th Century Political Economists

In hindsight, it is clear that the shift from the
domestic system to the factory system ranks as one
of the greatest achievements of modern civilization.
Specifically, it lifted what had, since the dawning

of the Neolithic era, been the main obstacle to
unlimited wealth, namely the muscular, human—or
animal—energy constraint.7 Never again would the
transformation of the earth's abundant natural
resources be constrained by muscular energy, or, for
that matter, unpredictable wind- or water-based
energy. More importantly, the yoke of two million
years of gruelling, physical, often times crippling
labor had been lifted. Output growth, it therefore
follows, would, from this point on, track energy
consumption growth, specifically, the growth of
"fire power" consumption. Clearly, the future looked
infinitely brighter.

How did 19th century scholars see this profound
change? What were their concerns, their doubts? How
did they formalize it? This section looks at the new
enabling technology that was the steam engine
(condensing, high-pressure and turbine) from the
point of view of 19th century political economists.
We will be particularly interested in addressing the
question of how and why? Specifically, how and why
did scholars take interest in this revolutionary
enabling technology.

As we shall argue, interest in the steam engine
was not spontaneous, but rather was the result of a
number of related issues and problems, the most
important of which were (i) the question of
financing the conversion of British industry to the
new technology and secondly (ii) once installed, the
perceived of inability of the British economy to
sustain high levels of output. Classical political
economists advocated free markets as a means of
reallocating Britain's wealth to the new technology.
Resources would have to be redirected to capital
investment, specifically in Watt-Boulton steam
engines. Mercantilism, they argued, was too rigid a
system to accomplish the task at hand.

Radical political economists focused their
attention on the shortcomings of the new gathering-
in system, specifically, with its perceived of
inability to generate sufficient demand to maintain
supply. Industrialist Robert Owen was one of the
first to raise the spectre of "failed transitions,"
which he attributed to underincome, namely, the
failure of income to increase commensurately with
productive capacity.

Early political economy, as such, should be
viewed as overwhelmingly normative in nature, having
little in common with present-day positive
economics. Absent was a need or desire to
accurately model production or distribution, let
alone exchange. For Smith, Ricardo, and "classical"
political economists in general, there was one
singular objective: moving to and staying on the
higher growth path defined by the new enabling

technology. The means were many, including free trade which, by reducing the price of foodstuffs at the turn of the century, would lead to lower real wages, thus increasing profits, savings and investment. The distribution of income, as Ricardo maintained, was the key issue for in it lay the key to the future of the U.K.'s capital stock, and, consequently, long-term growth. This section is organized chronologically and thematically. We begin with the early classical political economists. We then turn to the utopian socialists and radicals, and end with the neoclassical political economists.

The Early Classical Political Economists

How did early political economists fare? How did they see the changes that had been thrust upon Britain by the introduction of the steam engine? How did they approach this new energy form, and its myriad consequences? As we shall attempt to show here, they performed rather poorly regarding the former and admirably regarding the latter (i.e. the consequences). Energy was, for all intents and purposes, absent from their analysis. Foremost among the causes of this oversight was the absence of a scientific tradition in moral philosophy.8 As such, by contrast with contemporary political economy where modeling (formalizing) production processes is a goal in and of itself, classical political economists had no such interest. It sufficed to state that output was produced with labor and machinery, the former being an active factor input (source of value), and the latter being a passive one (Ricardo 1817).

That this was the case should not come as a surprise. After all, until then, there was no need for such a tradition. Production processes in 17th century Great Britain were Neolithic in scope. The means of creating wealth had gone unchanged for thousands of years, which explains, in large part, the absence of a "science of production" per se. Put otherwise, there was simply no need. Against this bucolic setting was thrust a new power source that increased productivity, altered the nature of work, and transformed the geographical localization of production (domestic system to factory system), thus altering forever the economic landscape. Also lacking was a theory of exchange and a theory of distribution.

Classical Political Economy

We begin with Adam Smith's An Inquiry into the
Nature and Causes of the Wealth of Nations which, we
maintain, is the first ever attempt at understanding
the changes thrust upon British society by the first
"energy revolution."[9] The division of labor,
according to Smith, had increased Great Britain's
potential to create wealth. The problem, however,
was with moving to the new, higher growth path. The
transition, according to Smith, was neither
automatic, nor instantaneous. The division of
labor, the ultimate source of wealth, was restricted
by the extent of the market. As we shall argue,
this is, in essence, the problem of underincome
referred to above.

In Book 1, Chapter 1, entitled "Of the Causes of
Improvement in the Productive Powers of Labour, and
of the Order according to which its Produce is
Naturally distributed among the different Ranks of
the People," he begins the "Wealth of Nations" with
an in-depth examination of the factors underlying
the increasing "productive powers of labor."
According to Smith, the principal factor is the
division of labor, broadly defined.

The greatest improvement in the productive powers
of labour, and the greatest part of the skill,
dexterity, and judgment with which it is anywhere
directed, or applied, seem to have been the effects
of the division of labour. (Smith 1776, 3)

The division of labor, he proceeds to argue, acts
upon labor productivity in three ways, namely in
terms of the "increased dexterity in every
particular workman," the "saving of the time which
is commonly lost in passing from one species of work
to another," and, lastly, to "the invention of a
great number of machines which facilitate and
abridge labour, and enable one man to do the work of
many" (Smith 1776, 5).

If, as we maintain, the "Wealth of Nations" is
about in the "energy revolution," then why is
machinery presented last? Why is energy or power
not so much as mentioned? The reasons are (i) the
state of the art of the science of heat in the late
18th century, (ii) the state of production theory at
this time and (iii) Smith's objectives. The first
two are self-evident. The science of heat was still
decades away, as was the very notion of "energy."
The best he could do, it therefore follows, was to
refer to power and machinery. Perhaps the most
important reason, however, is the underlying nature
of the "Wealth of Nations." Specifically, the
"Wealth of Nations" was not so much an attempt at
formalizing the "science of wealth" per se, as it
was an attempt to (i) convince the British public of

the great potential wealth made possible by machinery and, ultimately, by the steam engine, and (ii) alert it to the factors preventing it from reaching this "potential," including the notion that the "division of labor" is limited by the "extent of the market."

As argued earlier, the shift from the domestic system to the factory system altered both the underlying nature of production and the underlying nature of work. Specifically, inanimate energy replaced animate energy, and labor was reduced to a supervisory input (i.e. machine operatives). Labor, as such, was no longer physically productive, but, rather, was "organizationally productive." Theoretically speaking, machinery did not increase labor productivity, labor no longer being physically productive.11 Contrast this with Smith's view. Throughout the "Wealth of Nations," machinery is assumed to increase labor productivity. Titles such as "Of the Causes of Improvement in the Productive Powers of Labour, and the Order according to which its Produce is Naturally distributed among the different Ranks of the People," connote the notion that labor is more productive, when, in actual fact, it is no longer physically productive. we refer to this as the "laborcentricity" of classical production theory, and attribute it to decidedly Paleolithic nature of classical production theory. Classical political economists viewed machinery as "tools" when in actual fact, it was, in actual fact, a new energy source. Labor was becoming increasingly marginalized as a factor input.

This raises an important issue, namely intellectual fairness. Specifically, are these criticisms fair? Can we, or anyone else, be justified in criticizing Smith's pioneering work in the then embryonic "science of wealth?" The answer, in my view, is yes and no. No for the simple reason that Smith was operating in an intellectual vacuum as far as the "science of wealth" was concerned. He had little to work with from both natural and moral philosophy. On the other hand, such criticisms can be justified on what we choose to refer to as historical grounds. For example, in my view, it is useful to know that Smith's writings on production are theoretically and empirically inconsistent, if only to follow the course of subsequent developments. Clearly, by the end of the 18th century/beginning of the 19th century, it was imperative that inanimate power be included as a factor of production.

In the normal course of events, David Ricardo, with the benefit of 50 years of the factory system, would have improved, refined, corrected, and extended Smith's "positive" work on the "Nature and

Causes of the Wealth of Nations." Less emphasis would have been placed on the first two causes of the division of labor, and more on the third, namely machinery. By 1815, the Watts-Boulton rotary steam engine was commonplace in U.K. industry, and on the Continent. However, nowhere in The Principles of Political Economy and Taxation, published in 1817, does Ricardo mention either "fire power," "steam power," or "steam engine." In Chapter 20, entitled "Value and Riches, Their Distinctive Properties," he refers, albeit indirectly, to the effects of steam power on overall wealth.

> Value, then, essentially differs from riches, for value depends not on abundance, but on the difficulty or facility of production. The labour of a million of men in manufactures will always produce the same value, but will not always produce the same riches. By the invention of machinery, by improvements in skill, by a better division of labour, or by the discovery of new markets, where more advantageous exchanges may be made, a million of men may produce double, or treble the amount of riches, of "necessaries, conveniences and amusements," in one state of society that they could produce in another, but they will not, on that account, add anything to value. (Ricardo 1817, 182)

Judging by the numbers presented, it is clear that the effects of fire power on output were, in his eyes, not negligible. Fire power had lifted the yoke that was the energy constraint, a constraint which dated back to the Paleolithic era. Unfortunately, both he and Smith, not to mention other classical political economists, failed to identify by name fire power as the root cause of the sweeping changes in early 19th-century Great Britain. This, we maintain, had disastrous consequences, notably in the areas of classical value theory and in classical distribution theory. As if oblivious to the changing role of labor in production processes, Smith and Ricardo placed it at the center of classical value theory.[10] Accordingly, the greater the labor content, measured in hours, the higher the price. Clearly, from a strictly technical point of view, output in factories was increasing in the consumption of water and/or steam power, not in lower-level supervisors, which, as argued earlier, constitutes an organization input. It follows that, from the outset, classical political economy in general, and classical production theory in particular, were misspecified. Another unfortunate consequence was the machinery-labor productivity nexus that runs throughout the writings of Smith, Ricardo, and other classical political economists. According to historian Samuel Hollander, Smith viewed machinery as labor productivity increasing: "In a more precise sense, Smith recognized that certain refinements in the degree of specialization entail extensions in the period of production; for the invention of machines is portrayed as the outcome of the division of labor, and the use of machinery increases the roundaboutness of the process with the resultant increases in labour productivity" (Hollander 1987, 76). Here, the relevant issue is the definition of productivity. Steam power-driven machinery had rendered human, muscular energy redundant, transforming (reducing) labor into a supervisory input. Consequently, workers were no longer productive in the physical sense, but, instead, were productive in the organizational sense (See Section 1.3 in Chapter 1). That is, their presence was essential to the continued operation of the various machines.

Clearly, classical political economists failed to recognize these important changes. The question is why? Why did they fail to recognize and, consequently, formalize the new power drive technology that was steam? Why were they unable to see the parallels between animate, human, muscular energy and inanimate energy (despite the fact that the latter was measured in terms of the former (i.e. horsepower))? The answer, in my view, is relatively simple: classical political economy had, from the start, little to do with science as we know it today. Unlike today where knowledge of production, irrespective of policy, is an end in and of itself, in the late 18th/early 19th century, it was not. Put bluntly, there was no need for it. As such, only writings that addressed specific problems were deemed to be of any value (use?). The success of Smith and Ricardo as writers, we maintain, owes to the practical nature of the issues they chose to address. Foremost in Smith's mind were the obstacles to industrialization, notably the "extent of the market." Foremost in Ricardo's mind were the obstacles to on-going industrialization and growth, notably higher corn prices and wages, and higher taxes. The latter point is conveyed by the titles of his two most cited works, "Principles of Political Economy and Taxation," and "Essay on the Influence of a Low Price of Corn on the Profit of Stock."

This finding is consistent with Smith biographer Samuel Hollander's views on Smith's work.

> The main purpose of the Wealth of Nations was evidently not to provide an analytic framework

for its own sake. The object of the work was
ultimately to define the necessary conditions
for rapid economic development in contemporary
circumstances and Smith's treatment of the
price mechanism must accordingly, in the final
resort, be considered with this end in view.
(Hollander 1973, 307)

Classical Political Economy: Summary

As we have attempted to show here, classical
political economy, the product of the changes thrust
upon British society by the introduction of an
enabling technology in the form of the steam engine,
failed to explicitly incorporate energy in any form
in its analysis of the transformation of raw
materials and intermediate inputs into finished
products, thus precluding, in my view, any claim to
being scientific—in the current sense of the word.
Using a model of Paleolithic production processes,
they set out to provide analytical constructs for
what they saw as the true problem of political
economy, namely demand-constrained growth in the
case of Smith, and lower profits, savings,
investment, the capital stock, and, ultimately, the
rate of growth of the U.K. economy, in the case of
Ricardo.
The failure of classical political economy to
accurately model production in early 19th century
British—and European—manufacturing, we maintain,
contributed to its downfall and subsequent
usurpation by Marxian and neoclassical political
economy. While a power revolution in the form of
steam-powered machinery constituted the defining
force behind the phenomena they sought to
understand, it was, for all intents and purposes,
absent from their analysis. As we shall argue
below, these shortcomings (i.e. labor theory of
value) in combination with the worsening problem of
underincome, are what, in large measure, led to its
antithesis, namely Marxist political economy, and,
subsequently, to the classical counter-revolution,
namely neoclassical political economy. If labor was
the source of all value, as Smith and Ricardo
maintained, then, according to Marx, it only
followed that it be its sole claimant (i.e.
distribution wise). As such, Marxian political
economy should be understood as a one of the costs
of doing bad science. Further, neoclassical
political economy should be understood as a
refinement of classical political economy, one
intended to, among other things, render capital
productive. Paradoxically, at a time when energy
deepening was at its highest, the status of energy
as a factor input was at its lowest.

Robert Owen and Karl Marx

Judging by modern standards in science, classical
political economy as an attempt at understanding the
effects of the shift from the domestic system to the
factory system has to be considered a failure. That
this be the case is not altogether surprising given
the absence of a well-established scientific
tradition, and the heightened interest in policy
(e.g. maximizing growth, repealing the Corn Laws).
One could argue that they did not need—or at least
they thought—an empirically-consistent model of
production and exchange. It sufficed to point out
that output was produced by labor and machinery, the
former being paid wages and the latter profits; and
workers consumed most of their earnings while the
owners of capital saved most of theirs. Wages were,
in large measure, determined by the price of food.
Hence, higher food prices, like taxes on profits,
reduced savings, investment, and growth.

Value theory, we maintain, was a second-order
consideration, as evidenced by its primitive nature.
Goods, argued Smith and Ricardo, exchanged at ratios
determined by the relative amount of labor in each.
Just why they would choose labor as the relevant
standard at a time when it had been marginalized as
a factor input is anyone's guess? The reason, we
maintain, had little to do with logic, and
everything to do with convenience.[11] Value was not a
first-order consideration, nor was the problem of
exchange.

As a general rule, policy based on weak
fundamentals is, in the long run, doomed. This is
especially true when the policy in question involves
something as vital to the human condition as wages.
Classical political economists in general argued
that increasing profits would, in the long run, be
good for all of society, as more machinery would be
put in place, increasing productivity in all sectors
of the economy.[12] Profits were a social "good" as
they conferred benefits on all, including workers.
The process by which increased profits would lead to
higher savings, investment, growth, and ultimately,
a higher standard of living, was assumed to be
frictionless.   While relative inequities might
persist, absolute inequities would diminish as real
per-capita income increased (owing to a fall in
product prices).

The operative word here is "assumed." Reality
painted an altogether different picture of the new
industrial age. "Fire power" had revolutionized
manufacturing production processes, marginalized
workers, all-but-eliminated the various workers
guilds, and altered the way in which goods were

exchanged. First, as the data presented in Tables 2.5 and 2.6 show, steam power-induced increases in output, especially in the textiles sector, did not lead to higher nominal money income, but instead, to lower prices and, ultimately, to lower wages. In the face of either constant or rising food prices, workers' standard of living fell monotonically, as argued so eloquently by J. Fielden. Paradoxically, more productive workers (i.e. in the conventional sense) were worse off.

These developments, set against a theory of value based on labor, are, in my view, what fomented a number of "radical" movements in the 19th century. In short, the industrial revolution had passed labor by. Machinery decreased the demand for labor, all the while marginalizing it as a factor input, as evidenced by the increasing use of children in factories. Workers' standard of living had not increased, at least not commensurately with productivity. Moreover, the industrial system was, according to many, prone to crises of overproduction.

In this section, we examine the work of two such radicals, namely industrialist Robert Owen, and radical political economist, Karl Marx, two unlikely bedfellows, one being a factory owner, the other being an intellectual. It will be argued that their work should be understood as reactions to (i) the introduction of "fire power," (ii) the marginalization of labor (iii) "classical political economy" per se, and (iv) the falling relative price of manufactures—and hence, manufacturing wages—to food. Owen, a successful businessman, had witnessed the effects of the introduction of "scientific mechanical power" first hand, being the manager and, subsequently, the owner-manager of a number of textile mills, the most celebrated being at New Lanark, in Scotland. Consider, for example, the following passage taken from the Report to the County of Lanark, published in 1820:

> It is well known that, during the last half century in particular, Great Britain, beyond any other nation, has progressively increased its powers of production, by a rapid advancement in scientific improvements and arrangements, introduced, more or less, into all the departments of productive industry throughout the empire.
> The amount of this new productive power cannot, for want of proper data, be very accurately estimated; but your Reporter has ascertained from facts which none will dispute, that its increase has been enormous; that, compared with the manual labour of the whole population of Great Britain and Ireland, it is,

at least, as forty to one, and may be easily
made as 100 to one; and that this increase may
be extended to other countries; that it is
already sufficient to saturate the world with
wealth and that the power of creating wealth
may be made to advance perpetually in an
accelerating ratio. (Owen 1820, 246)

That Owen, more so than either Smith or Ricardo,
identified and analyzed the changes thrust upon
society by "fire power," can be attributed, in large
measure, to what, for lack of a better word, is his
life experience, specifically, to having been
exposed, as early as the late 1780's, to water and
steam power.   After having, with John Jones,
manufactured and sold roving and machinery to the
spinning industry, he went to work for Peter
Drinkwater, the first man to build a mill in
Manchester powered by a Watt rotary steam engine
(Butt 1971, 169).
Throughout his writings, Owen refers, often times
in glowing terms, to the energy-based innovation
which was the introduction of steam power-based
drive in British manufacturing. Not knowing just how
to describe it, he refers to it both as "mechanical"
and "chemical" power as the following passage bears
witness:

It must be admitted that scientific or
artificial aid to man increases his productive
powers, his natural wants remaining the same;
and in proportion as his productive powers
increase he becomes less dependent on his
physical strength and on the many contingencies
connected with it.... That the direct effect
of every addition to scientific, or mechanical
and chemical power is to increase wealth; and
it is found, accordingly, that the immediate
cause of the present want of employment from
the working classes is an excess of production
of all kinds of wealth, by which, under the
existing arrangements of commerce, all the
markets of the world are overstocked. (Owen
1820, 247)

Unlike the classical political economists, Owen
was acutely aware of the consequences of "scientific
power" for labor. As we pointed out in Chapter 2,
it had transformed labor into a form of lower-level
supervisory input, overseeing the workings of steam-
powered machine. He referred to this as "mental
power." In short, Owen was the first "energy
economist," being the first to describe in
considerable detail the many changes thrust upon
British society by the introduction of a new energy
source, namely "scientific power."

In many ways, Owen had much in common with Smith and Ricardo. Take, for example, the question of motives. All sought to promote growth based on the newly-discovered source of inanimate power. Where they differed, however, is with regard to the relevant "means." Smith and Ricardo held that lower wages, by increasing profits, would increase savings and investment, and ultimately, growth. Owen saw things differently. Whereas Ricardo held that productive capacity (i.e. supply) creates a proportionate level of income, and that this income creates a proportionate level of demand, Owen disagreed, largely based on his experience with "scientific power" in industry. In short, Owen had identified the problem of underincome. Given that profits are a residual form of money income, an increase in output, due to "scientific power" did not result in an increase in money income, and consequently, "consumption" as expenditure was then known. In fact, argued Owen, wages had fallen, exacerbating the problem of underincome.

Owen was acutely aware that an increase in output, due to "scientific power," did not increase money income. "Prime costs," as he referred to them, did not increase with output, making for underincome. The move to the factory system, as pointed out above, had altered the underlying nature of exchange. Textile producers in Manchester and elsewhere financed their variable costs "Prime costs" with trade credit. As we shall argue here, this flaw in the newly instituted factory system (i.e. partial cash-in-advance constraint contributing to underincome) was the motivating factor underlying and running through most of his writings. "Scientific power" had increased productive capacity and productivity, yet workers and society as a whole were worse off. Consider, for example, the closing passage of his Report to the County of Lanark:

> Your Reporter solicits no favour from any party; he belongs to none. He merely calls upon those who are the most competent to the task, honestly, as men valuing their own interests and the interests of society, to investigate, without favour or affection, a "Plan (derived from thirty years' study and practical experience) for relieving public distress and removing discontent, by giving permanent productive employment to the poor and working classes, under arrangements which will essentially improve their character and ameliorate their condition, diminish the expenses of production and consumption, and create markets coextensive with production."
> (Owen 1820, 298)

The problem of underincome runs throughout the Report to the County of Lanark. Consider, for example, the preamble.

> Having taken this view of the subject, your Reporter was induced to conclude that the want of beneficial employment for the working classes, and the consequent public distress, were owing to the rapid increase of the new productive power, for the advantageous application of which, society had neglected to make the proper arrangements. Could these arrangements be formed, he entertained the most confident expectation that productive employment might again be found for all who required it; and that the national distress, of which all now so loudly complain, might be gradually converted into a much higher degree of prosperity than was attainable prior to the extraordinary accession lately made to the productive powers of society.
>
> Cheered by such a prospect, your Reporter directed his attention to the consideration of the possibility of devising arrangements by means of which the whole population might participate in the benefits derivable from the increase of scientific productive power; and has the satisfaction to state to the meeting, that he has strong grounds to believe that such arrangements are practicable.
>
> His opinion on this important part of the subject is founded on the following considerations:
>
> First.—It must be admitted that scientific or artificial aid to man increases his productive powers, his natural wants remaining the same; and in proportion as his productive powers increase he becomes less dependent on his physical strength and on the many contingencies connected with it.
>
> Second. That the direct effect of every addition to scientific or mechanical and chemical power is to increase wealth; and it is found, accordingly, that the immediate cause of the present want of employment for the working classes is an excess of production of all kinds of wealth, by which, under the existing arrangements of commerce, all the markets of the world are overstocked.
>
> Third. That, could markets be found, an incalculable addition might yet be made to the wealth of society, as is most evident from the number of persons who seek employment, and the far greater number who, from ignorance, are inefficiently employed, but still more from the means we possess of increasing, to an unlimited extent, our scientific powers of production.
>
> Fourth.—That the deficiency of employment for the working classes cannot proceed from a want of wealth or capital, or of the means of greatly adding to that which now exists, but

> from some defect in the mode of distributing
> this extraordinary addition of new capital
> throughout society, or, to speak commercially,
> from the want of a market, or means of
> exchange, co-extensive with the means of
> production. (Owen 1820, 248)

To address these problems, Owen, in his report to
the Committee of Gentlemen of the Upper Ward of
Lanarkshire, proposed a "radical" reorganization of
distribution, and, to a lesser extent, production,
both for agriculture and industry. We shall focus
on the latter. The core change, proposed by Owen,
is with the standard of value. Specifically, he
proposed replacing what is an artificial standard of
value (i.e. money) by a natural one, namely human
labor, which he refers to as the "combined manual
and mental powers of men called into action." In
actual fact, what Owen has in mind is an energy-
based theory of value, based on labor's manual and
mental power.

> Already, however, the average physical power of
> men as well as of horses (equally varied in
> individuals), has been calculated for
> scientific purposes, and both now serve to
> measure inanimate powers. One the same
> principle the average of human labour or power
> may be ascertained; and as it forms the essence
> of all wealth, its value in every article of
> produce may also be ascertained, and its
> exchangeable value with all other goods fixed
> accordingly; the whole to be permanent for a
> given period. (Owen 1820, 251)

This was the first-step in Owen's plan for
"distribution." Scientific power, in combination
with lower nominal wages, had reduced the cost of
labor (i.e. supervision) per unit output. To
reverse this trend, Owen argued in favor of a labor
standard of value in which the amount of labor per
unit output would be determined and held fixed (i.e.
for a given period). Any increase in output would,
as such, automatically lead to an increase in wages,
thus solving, at least in part, the problem of
underincome. Income would increase commensurately
with output. Markets would expand as a result,
leading to increased profits. In short, a win-win
situation.
From a purely scientific point of view, this
plan, while feasible, was inconsistent with the
facts. As pointed out above, "scientific power" had
marginalized labor, making it "child's play," so to
speak. The labor standard, as Owen referred to it,
was orthogonal to the technological reality of the
early 19th century. How, then, should Owen's plan

be understood? One could argue that the labor standard had more to do with the problem of exchange than with the problem of production. By stating, prima facie, that labor was the source of all value, Owen could then mount a case for higher wages, consumption, and, consequently, profits. Without it, there would be nothing, owing, in large measure, to the residual nature of profits in the factory system.

Consider the following passage in which he describes such a system.

> The genuine principle of barter was, to exchange the supposed prime cost of, or value of labour in, one article, against the prime cost of, or value of labour contained in any other article. This is the only equitable principle of exchange; but, as inventions increased and human desires multiplied, it was found to be inconvenient in practice. Barter was succeeded by commerce, the principle of which is, to produce or procure every article at the lowest, and to obtain for it, in exchange, the highest amount of labour. To effect this, an artificial standard of value was necessary; and metals were, by common consent among nations, permitted to perform the office.

> This principle, in the progress of its operation, has been productive of important advantages, and of very great evils; but, like barter, it has been suited to a certain stage of society. It has stimulated invention; it has given industry and talent to the human character; and has secured the future exertion of these energies which otherwise might have remained dormant and unknown.

> But it has made man ignorantly, individually selfish; placed him in opposition to his fellows; engendered fraud and deceit; blindly urged him forward to create, but deprived him of the wisdom to enjoy. In striving to take advantage of others, he has over-reached himself. The strong hand of necessity will now force him into the path which conducts to that wisdom in which he has been so long deficient. He will discover the advantages to be derived from uniting in practice the best parts of the principles of barter and commerce, and dismissing those which experience has proved to be inconvenient and injurious. (Owen 1820, 263)

In short, according to Owen, the introduction of "scientific power" had shaken the foundations of 18th-century British society. Potential output had increased manifold, yet welfare had failed to increase. In fact, according to Owen, things had deteriorated. A new system was needed. The Report

to the County of Lanark should be seen as Owen's
blueprint for a better tomorrow, one where
scientific power is put to use to better mankind's
material and intellectual existence.

Let me now turn to German political economist,
Karl Marx, who, like Owen, is both scientist and
ideologue. In Chapters 14 and 15 of Das Capital,
published in 1867, he presented what we maintain is
the most scientifically accurate description of
production processes both before and after the
industrial revolution. Consider, for example, the
following passage which describes the role of tools
and power in "heterogeneous" and "serial"
manufactures, the former referring to the domestic
system, and the latter, to the factory system.

> Mathematicians and mechanicians, and in this
> they are followed by a few English economists,
> call a tool a simple machine, and a machine a
> complex tool. They see no essential difference
> between them, and even give the name of machine
> to the simple mechanical powers, the lever, the
> inclined plane, the screw, the wedge, etc. As
> a matter of fact, every machine is a
> combination of those simple powers, no matter
> how they may be disguised. From the economic
> standpoint, this explanation is worth nothing,
> because the historical element is wanting.
> Another explanation of the difference between
> tool and machine is that, in the case of the
> tool, man is the motive power, while the motive
> power of a machine is something different from
> man, is, for instance, an animal, water, wind,
> and so on. According to this, a plough drawn
> by oxen, which is a contrivance common to the
> most different epochs, would be a machine,
> while Claussen's circular loom, which, worked
> by a single labourer, weaves 96,000 picks per
> minute, would be a mere tool. Nay, this very
> loom, though a tool when worked by hand, would,
> if worked by steam, be a machine. And, since
> the application of animal power is one of man's
> earliest inventions, production by machinery
> would have preceded production by handicrafts.
> When in 1735, John Wyalt brought out his
> spinning machine and began the industrial
> revolution of the eighteenth century, not a
> word did he say about an ass driving it instead
> of a man, and yet this part fell to the ass.
> He described it as a machine "to spin without
> fingers."
> All fully developed machinery consists of
> three essentially different parts, the motor
> mechanism, the transmitting mechanism, and
> finally the tool of working machine. The motor
> mechanism is that which puts the whole in
> motion. It either generates its own motive
> power, like the steam engine, the caloric
> engine, the electro-magnetic machine, etc., or
> it receives its impulse from some already

> existing natural force, like the water-wheel
> from a head of water, the windmill from wind,
> etc. The transmitting mechanism, composed of
> flywheels, shafting, cogwheels, pulleys,
> straps, ropes, bands, pinions, and gearing of
> the most varied kinds, regulates the motion,
> changes its form where necessary, as, for
> instance, from linear to circular, and divides
> and distributes it among the working machines.
> These two parts of the whole mechanism are
> there solely for putting the working machines
> in motion, by means of which motion the subject
> of labour is seized upon and modified as
> desired. The tool or working machine is that
> part of machinery with which the industrial
> revolution of the eighteenth century started.
> And, to this day it constantly serves as such a
> starting point whenever a handicraft, or a
> manufacture, is turned into industry carried on
> by machinery. (Marx 1991, 181)

What is particularly interesting—and, to a
certain extent, puzzling—is his decision to place
these two chapters (Chapters 14 and 15) after the
bulk of his substantive analysis, notably on
exchange, surplus value, exploitation, etcetera
(Chapters 1-13). Equally puzzling is the fact that
in Chapters 1 13, labor is the source of all value.
That is, labor is the only productive factor. How
then are we to understand what, by all accounts,
appears to be an egregious error? Specifically,
Marx was well aware of the fact that labor had been
transformed from a source of energy and
organization, to a source of organization only, yet
in the crux of his analysis, he focuses exclusively
on labor.

The answer, we believe, lies in the duality
originally found in Owen's work. Fifty years
earlier, Owen knew full well that labor had become a
minor input in the factory system, yet, when the
time came to examine the problem of distribution,
put it at the center of his analysis. The reason,
as argued earlier, has little to do with science,
and everything to do with "ideology." Labor, Owen
felt, should receive a larger share of energy rents.
Marx held similar views, which, we maintain,
explains the subsidiary nature of his work on
machines and power. It also delineates, quite well,
the duality of Marx as both a scientist, and
ideologue.

The next question relates to the problem of
exchange and the problem of underincome.
Specifically, was Marx aware of the problem of
exchange and the problem of underincome? Despite
devoting the first three (3) chapters of Das Capital
to money and exchange, he fails to describe, in a
reasonably accurate way, the process of exchange in

industrialized economies. Absent from his analysis is any mention of the residual nature of profits (surplus value), and its consequences at the aggregate level. In short, he assumes, like Ricardo, that the value of money is identically equal to the value of output. Consequently, the problem of underincome is ignored.

The last question relates to the problem of distribution. As we have argued (Beaudreau 1998), Das Capital should be seen not so much as an attempt at understanding production, exchange and distribution, as a conscientious attempt, by a perceptive individual, of positively influencing labor's bargaining power in the relevant energy-rent bargaining problem. As the data presented above reveal, factory workers in mid-19th century Great Britain suffered greatly from falling textile prices and wages. The riches of the power revolution, so to speak, had passed them by. They were no better off in the mid 1800's than they were in the early 1800's. Marx's theory of distribution, based on labor as the sole source of value, should, as such, be seen for what it was: an attempt at influencing the parameters of the relevant bargaining problem. If labor could reasonably (read: scientifically) claim that all value was rightfully theirs, then wages would increase, and consequently, so would labor's standard of living. As it turns out, this was also the preferred approach of Robert Owen at the turn of the century, and, interestingly enough, was not inconsistent with the labor theory of value.

The Neoclassical Political Economists

By the mid-19th century, a new threat to savings, investment and steam engine-based growth, the staples of classical political economy, had emerged in Europe, namely radical political economy, defined broadly to include Marxian political economy, glut theory and underconsumption theory. Like Smith's "extent of the market," and Ricardo's "high food prices," numerous social, political and economic radicals now threatened the existing order. Whereas critics like Robert Owen sought to reform the factory system, this new generation of reformers aimed at nothing less than its end, to be replaced by a new, social and economic order built around the worker.

As we shall argue here, this new threat was, in part, self-inflicted. Specifically, the classical political economists' failure to accurately model production processes, in combination with the problem of underincome, were, in large measure, responsible for the mercurial rise of radical political economy. Robert Owen and Karl Marx's basic ideological postulate (axiom) was the labor theory of value in its simplest form. Labor was the source of all value (Smith 1776; Ricardo 1817). Ergo, something must be askew, they reasoned, as labor income was on the wane. Logically speaking, they were right. If, as the classical political economists maintained, labor was the source of all value, then to deprive it of its product (marginal revenue product) was akin to theft.

The classical political economists were, for the most part, uninterested in production, beyond the obvious. Labor and tools had, were, and, undoubtedly, would continue to transform raw materials into valuable goods. Their principal preoccupation was with distribution, putting an inordinate amount of attention on the relationship between the price of corn, wages, profits, and ultimately growth. As we shall argue here, their ideological successors, the neoclassical political economists, did little to set the record straight. Instead of providing an empirically-consistent view of production based on then well-established science of thermodynamics (i.e. energy) and organization, they chose to (i) develop a non-substantive theory of value (utility-based) and (ii) add tools to the list of productive inputs (Mirowski 1990). Both, it bears noting, borrowed, analytically-speaking, from thermodynamics.

Once again, the problem of distribution took center stage, but for altogether different reasons, reasons which, at this point, were purely defensive in nature. Higher corn prices had threatened profit growth, savings growth, investment growth, and overall growth at the turn of the century. Falling textile prices and wages in the textile industry threatening profit growth, savings growth, investment growth, and overall growth. Now, socialism and communism threatened the very cornerstone of capitalism, namely the institution of private property. Consider the following passage, taken from the Preface to the first edition of William Stanley Jevons' The Theory of Political Economy, published, it bears noting, in 1871, four years after Karl Marx's Das Capital.

> There are many portions of the economical doctrine which appear to me as scientific in form as they are consonant with facts. We would especially mention the theories of population and of rent.... Other generally-accepted doctrines have always appeared to me purely delusive, especially, the so-called wage fund theory. This theory pretends to give a solution to the main problem of the science—to determine the wages of labour—yet on close

> examination, its conclusion is a mere truism,
> namely that the average rate of wages is found
> by dividing the whole amount appropriated to
> the payment of wages by the number of those
> between whom it is divided. (Jevons 1871, 1)

Roughly a century after Adam Smith's pathbreaking contribution, the problem of value and distribution had not been resolved, at least, not to everyone's satisfaction. In hindsight, Jevons was right: classical political economy was woefully inadequate. Labor, after all, was, by now, a marginal factor, supervising machinery (machine operatives).[13] The relevant question, it follows, is whether Jevons, Marshall and Walras were successful in their attempt to provide a scientific theory of value and distribution?

I maintain that they were not, owing mainly to their failure to provide an empirically-correct model of production. Neoclassical political economists, by moving value theory away from "substances," widened the chasm which separated theory from reality. In lieu of a theory of value based on energy as a factor of production (i.e. à la Owen), they offered a theory of value based on energy as a consumption good (Edgeworth 1889; Mirowski 1988,1993) inspired largely from the then-nascent science of thermodynamics.

Like their classical forefathers, Jevons, Marshall and Walras continued to model production as a "Paleolithic" activity involving labor and capital. Energy is conspicuous for its absence. Take, for example, the description of production in the cotton industry found in Chapter VII of Jevons' The Theory of Political Economy, entitled "Theory of Capital".

> The ultimate object of all industry engaged in
> cotton is the production of cotton goods. But
> the complete process of producing those goods
> is divided into many parts; and it is necessary
> to begin the spending of labor a long time
> before any goods can be finished.
> In the first place, labour will be required
> to till the land which is to beat the cotton
> plants, and probably two years at least will
> elapse between the time when the ground is
> first broken and the time when the cotton
> reaches the mills. A cotton mill, again, must
> be a very strong and durable structure, and
> must contain machinery of a very costly
> character, which can only repay its owner by a
> long course of use. We might spin and weave
> cotton goods as in former times, or as it is
> done in Cashmere, with a very small use of
> capital; but then the labour required would be
> enormously greater in proportion to the

produce. It is far more economical to the end
to spend a vast amount of labour and capital
building a substantial mill and filling it with
the best machinery, which will then go on
working with unimpaired efficiency for thirty
years or more. This means that, in addition to
the labour spent in superintending the machines
at the moment when the goods are produced, a
great quantity of labor has been spent from one
to thirty years in advance. (Jevons 1871, 229)

The complete absence of energy from The Theory of
Political Economy is all the more puzzling in light
of Jevons' earlier work The Coal Question, published
six years earlier, in 1865.   Consider, for example,
the opening paragraph of Chapter 1, entitled
Introduction and Outline.[14]

Day by Day it becomes more evident that the
Coal we happily possess in excellent quality
and abundance is the mainspring of modern
material civilization. As the source of fire,
it is the source at once of mechanical motion
and of chemical change. Accordingly it is the
chief agent in almost every improvement or
discovery in the arts which the present age
brings forth.... And as the source especially
of steam and iron, coal is all powerful. This
age has been called the Iron Age, and it is
true that iron is the material of most great
novelties.   By its strength, endurance, and
wide range of qualities, this metal is fitted
to the fulcrum and lever of great works, while
steam is the motive power. But coal alone can
command in sufficient abundance either the iron
or the steam; and coal, therefore, commands
this age—the Age of Coal.
      Coal in truth stands not beside, but
entirely above all other commodities, It is
the material source of the energy of the
country—the universal aid—the factor in
everything we do. With coal almost any feat is
possible or easy; without it we are thrown back
into the laborious poverty of early times.
With such facts familiarly before us, it can be
no matter of surprise that year by year we make
larger draughts upon a material of such myriad
qualities—of such miraculous powers. But it
is at the same time impossible that men of
foresight should not turn to compare with some
anxiety the masses yearly drawn with the
quantities known or supposed to lies within
these islands. (Jevons 1965, 2)

Next, take Chapter VII of Marshall's Principles
of Economics, entitled "The Growth of Wealth," where
the emphasis, despite a historical reference to
water and steam power, is on capital.

The implements of the English farmer had been
rising slowly in value for a long time; but the
progress was quickened in the eighteenth
century. After a while the use first of water
power and then of steam power caused the rapid
substitution of expensive machinery for
inexpensive hand tools in one department of
production after another. As in earlier times
the most expensive implements were ships and in
some cases canals for navigation and
irrigation, so now they are the means of
locomotion in general;-railways and tramways,
canals, docks and ships, telegraph and
telephone systems and water-works; even gas-
works might almost come under this head, on the
ground that al great part of their plant is
devoted to distributing the gas. After these
come mines and iron and chemical works, ship-
building years, printing-presses, and other
large factories full of machinery.
On whichever side we look we find that the
progress and diffusion of knowledge are
constantly leading to the adoption of new
processes and new machinery which economise on
human effort on condition that some of the
effort is spent a good while before the
attainment of the ultimate ends to which it is
directed. (Marshall 1890, 184)

Conceptually, capital predicates the "growth of
wealth."

The whole history of man shows that his wants
expanded with the growth of wealth and
knowledge. And with the growth of openings for
investment of capital there is a constant
increase in that surplus of production over the
necessaries of life, which gives the power to
save. When the arts of production were rude,
there was very little surplus, except when a
strong ruling race kept the subject masses hard
at work on the bare necessaries of life, and
where the climate was so mild that those
necessaries were small and easily obtained.
But ever increase in the arts of production,
and in the capital accumulated to assist and
support labour in future production, increased
the surplus out of which more wealth could be
accumulated. After a time civilization became
possible in temperate and even in cold
climates; the increase of material wealth was
possible under conditions that did not enervate
the worker, and did not therefore destroy the
foundations in which it rested. Thus, from
step to step wealth and knowledge have grown,
and with every step the power of saving wealth
and extending knowledge has increased.
(Marshall 1890, 186)

Analytically, capital, unproductive in classical
political economy, is simply decreed to be

productive. Borrowing from the methodology of thermodynamics, specifically from second-law analysis where efficiency is increasing in the design (quality) of the capital equipment, neoclassical writers simply assumed rising total productivity and declining marginal productivity for both capital and labor. Factors would, from now on, be infinitely substitutable, at least in theory. Most importantly, capital was now physically productive. Heretofore, tools would be productive, in direct violation of classical mechanics where tools "transmit force or torque." This, however, provided the much-needed theoretical basis for a model of labor market and capital-market behavior based on the notion of marginal productivity (i.e. the demand for labor and capital).

Why was energy ignored in neoclassical political economy, especially at a time when it was the toast of the natural sciences? The reasons, we believe, are numerous. First, there is the question of objectives. Neoclassical political economy, as we have maintained here, was, from the start, defensive in nature. In fact, one could go as far as to argue that were it not for the problem of underincome, we would, to this day, be classical political economists. A second reason is the absence of energy deepening in the late 19th century. By 1860, most of industry in Great Britain had converted to high-pressure steam (as opposed to atmospheric (condensing)) as the chief power source driving continuous-flow machinery. Hence, there was nothing new on the energy front—that is, to report by 1870. One could argue counterfactually that had further energy deepening occurred in this period, then the chances of seeing energy in the production function would have been greater—perhaps not much, however.

Not including energy in the production function had important implications for the problem of distribution as seen by neoclassical political economists. Specifically, the notion of energy rents was unknown to them. Now that both capital and labor were deemed to be physically productive, the problem of distribution consisted of measuring their physical contribution to output and remunerating accordingly. This is now known as the neoclassical theory of distribution.

From a purely counterfactual point of view, one cannot help but wonder what could have been had Jevons, Marshall and Walras probed more deeply into the inner workings of production processes. Instead of being a mere cog in a pseudo-scientific model of distribution, thermodynamics, the science of energy, could have, at least conceivably, become the cornerstone of production theory, and, consequently, distribution and consumption theory. There,

political economists would have found the key to
understanding the creation and "growth" of wealth,
not to mention the many changes which had
characterized the 19th century.   The "Energetics"
movement, founded by physicists such as Hermann Von
Helmholtz, was, at this time, moving in this very
direction. According to sociologist Ansom Rabinbach:

> As Helmholtz was aware, the breakthrough in
> thermodynamics   had   enormous   social
> implications.   In his popular lectures and
> writings, he strikingly portrayed the movements
> of the planets, the forces of nature, the
> productive forces of machines, and, of course,
> human labor power as examples of the principle
> of the conservation of energy. The cosmos was
> essentially   a   system   of   production   whose
> product was the universal Kraft, necessary to
> power the engines of nature and society, a vast
> and protean reservoir of labor power awaiting
> its conversion to work. (Rabinbach 1990, 3)

Why the likes of Edgeworth, Jevons, Marshall,
Walras and others chose to ignore this development
is a topic for future research.[15]

Nineteenth   Century   Britain   as   Seen   by Contemporary Political Economists

This chapter has been critical of 19th century
political economy, from Smith to Owen to Marx to
Marshall.   Chief among the criticisms has been the
little   attention—in   fact,   the   total   lack   of
attention—paid to energy in their work.   Output is
modeled in Paleolithic terms, increasing in capital
and labor.   This oversight, we maintain, lies at the
root of what we refer to as the fall of political
economy.   That is, energy in general, and energy
innovations   in   particular   were   the   cause   of
political economy, yet, they were nowhere to be
found.
Unfortunately, the passage of time has failed to
correct the problem.   More than two centuries after
the first industrial revolution and a century after
the development of neoclassical political economy,
energy remains the forgotten input.   In recent work
on the energy and the second industrial revolution
(i.e. electrification) (Beaudreau 1995,1996a,1998),
we alluded to the existence, in the literature, of a
dichotomy regarding energy in general and electric
power in particular, namely of the historical record
pointing to the important role of energy in general
and electric power in particular in U.S. economic
growth, and of the analytical growth theory where
they are absent altogether.

> The bulk of the historical evidence points to
> energy in general, and electric power in
> particular, to be an important cause of
> productivity and output growth, yet, on the
> other hand, analytical studies of the sources
> of growth find it to be marginally important.
> Thus, either the historical record is
> incorrect, or previous studies have
> underestimated the role of energy in general
> and electric power in particular in the process
> of economic growth. (Beaudreau 1995, 232)

Ironically, but not altogether surprisingly, a
similar dichotomy also characterizes the literature
on steam power and economic growth in the 19th
century. The irony in this case stems from the
absence in analytical work of what most consider to
be the root cause of the industrial revolution,
namely the steam engine. A good of example of this
is a recent compendium on British economic growth
since 1700, edited by Roderick Floud and Donald
McCloskey (Floud and McCloskey 1994) in which
contributors make ample reference is made to energy,
energy utilization, steam power, fire power, thermal
and kinetic energy, coal consumption, etcetera, but
stop short of including energy as a factor of
production. Joel Mokyr, for example, in an article
entitled "Technological change, 1700-1830," devotes
a section to "Energy Utilization," where he
describes in some detail the shift from hydraulic
energy to thermal energy, without relating energy
and productivity. Nick Crafts, in an article
entitled "The Industrial Revolution," performs
standard growth accounting on the British economy
for the period 1700-1860, and finds Solow residuals"
which vary in importance from 10 percent for the
period 1760-1801, to 28 percent for 1700-1760. The
relevant "residual" for the period 1801-1831 is 18
percent. Again, energy is not included as a factor
input.

Donald McCloskey, in an article entitled "1780-
1860: A Survey," which, in a strange way, bears a
striking resemblance to the current literature on
the "Productivity Paradox" and the "Productivity
Slowdown," cuts to the chase. Pointing to data
which show that real income per head "nowadays" is
twelve times greater than in 1780, he maintains that
political economy in general, and economic
historians in particular, have failed in their
attempts at explaining these trends.

> The conclusion, then, is that Harberger
> triangles—which is to say the gains from
> efficiency at the margin—cannot explain the
> factor of twelve. This is lamentable, because
> economics is much more confident about static

arguments   than   about   dynamic   arguments.
(McCloskey 1994, 270)

While McCloskey offers a number of promising
research avenues such as "the role of persuasive
talk in the economy," we maintain, in keeping with
the main thesis of this book, that the essential
problem with the literature on British economic
growth in the 18th and 19th centuries is with the
relevant model of production, specifically, with the
absence of energy in the various analytical models.
Ironically,   while   every   economic   historian—bar
none—agrees that thermal energy in the form of steam
power is what launched the industrial revolution,
energy as a factor—of production—is absent from
work on 19th century economic growth. A.E. Musson, in
Industrial Motive Power in the United Kingdom, 1800-
70, makes a similar point:

> It   is   generally   recognized   that   the
> introduction of steam power was a crucial
> factor in the Industrial Revolution, closely
> linked with the exploitation of Britain's coal
> and   iron   resources,   the   development   of
> mechanical engineering, and the growth of the
> factory system.   A considerable amount of
> historical research has been carried out into
> the   scientific-technological   developments
> involved—into   how   the   steam   engine   was
> developed from the crude creations of Savery
> and   Newcommen   to   the   comparatively
> sophisticated products of Watt's genius, with
> separate condenser, air pump, direct action,
> rotative motion, etc.—and how steam power began
> to spread into coal-mining, cotton spinning,
> flour milling, brewing, and various other
> industries in the eighteenth century.... In
> contrast   to   the   interest   in   the   early
> pioneering, there has been comparatively little
> effort to investigate the massive growth and
> spread of steam power after 1800, except in a
> few industries and areas. Several factors have
> contributed to this neglect.   The "heroic"
> theory of historical evolution has tended to
> concentrate attention primarily on Newcommen
> and Watt, though with some interest in
> Trevithick   and   Woolf   for   their   early
> development of high-pressure engines. (Musson
> 1974, 415)

This raises other questions. Why, for example,
have historians concentrated on Newcommen and Watt,
ignoring Trevithick, Woolf and Parsons? The answer,
we believe, is simple, namely the absence of energy
in the theory of production from Adam Smith to Paul
Romer. Historically, energy has found its way into
production via the technology coefficient (i.e. the

A(t) scaler in the Cobb-Douglas production function)(Jorgenson 1983; Bresnahan and Trajtenberg 1992; Helpman and Trajtenberg 1994). Since shocks occur, by design, at points in time, and not over time (i.e. a century), it stands to reason that energy innovations would be studied as distinct, one-shot events.[16]

The consequences of this "neglect," Musson points out were real:

> This neglect, however, has had serious consequences for the historical interpretation of Britain's nineteenth-century industrial development. The overwhelming emphasis on the initial growth of steam power has tended to create a very misleading view of the pace and extent of early steam-powered mechanization in British industry. (Musson 1974, 416)

Conclusions

This chapter has "deconstructed" the first industrial revolution into its component parts and component sub-periods. As was shown, the first industrial revolutions were about power, specifically about energy deepening. It was not about James Watt's steam engine, but rather about the power consumption and work output that resulted. As such, the steam engine was not the cause of the massive increase in material wealth that resulted. Rather, it acted as an enabling technology, providing the framework for a century of energy deepening. The latter was achieved in three sub-periods, notably the early 19th century, the mid-19th century and the late 19th century. Each of these was characterized by a distinct steam power-drive technology, going from Watt's early engine, to high-pressure steam engines, to steam turbines.

As was shown, this deconstruction had not escaped the attention of a number of observers/scholars, notably Karl Marx who understood, perhaps more than any other classical political economist, the technological facilitator-energy consumption distinction. For someone who attributed the creation of all material wealth to labor, Marx was acutely aware of the fact that power/energy is what lay at the root of the first industrial revolution.

To summarize, the first industrial revolution was about two things, an enabling technology in the form of three generations of steam engines (Watt-Boulton, High-Pressure and Turbine) and an energy source, in this case, coal. As such, it was not a one-shot occurrence, but rather was a long, drawn out one. Energy deepening continued throughout the 19th

century, resulting in high levels of conventionally-
defined labor productivity growth $(W(t)/S(t))$.

3

The Second Industrial
Revolution Deconstructed

> The speed with which electricity was adopted
> may be readily indicated. Electric motors
> accounted for less than 5 percent of total
> installed horsepower in American manufacturing
> in 1899. The growth in the first years of the
> twentieth century was such that by 1909 their
> share of manufacturing horsepower was 25
> percent. Ten years later the share rose to 55
> percent and by 1929 electric motors completely
> dominated the manufacturing sector by providing
> over 80 percent of total installed horsepower.
> The sharp rise in productivity in the American
> economy, in the years after World War I,
> doubtless owed a great deal, both directly and
> indirectly, to the electrification of
> manufacturing.
>
> —Nathan Rosenberg, Technology and
> American Economic Growth

Introduction

The shift from the domestic to the factory system
had lifted the energy constraint which had, for over
two million years, constrained Homo sapien's
(neanderthalensis and sapiens) ability to transform
the earth's abundant supply of raw materials into
goods and services. The constraint, however, had
not been lifted completely, owing in large measure
to the associated power transmission technology,
namely the cumbersome belting, gearing and shafting.
Further, not all production processes at the time
were amenable to mechanization (i.e. driven by steam
power). Part of this owed to the size of
economically-viable steam engines (especially high-
pressure engines, steam turbines), limiting
potential applications to large concerns.

Enter a new power transmission technology (enabling technology), namely, electro-magnetic motors (hereafter, electric motors). Unlike steam engines, electric motors came in different sizes and more importantly, different—read, faster—speeds. As they began appearing in factories (in replacement of steam engines), the energy constraint was, once again, lifted and pushed back, in this case, further than any ever before. Production processes that had, hitherto, not been mechanized were. Henry Ford's electric motor-driven assembly line is a case in point. Static material-handling processes of which the final assembly of the automobile is a case in point, (i.e. the craft system) were replaced with continuous-flow materials handling processes, driven by direct-current (DC) electric motors. Secondly, production processes that were mechanized (i.e. powered by steam) were converted to electric power, resulting in even greater speeds (throughput rates), and consequently, higher throughput and productivity. The result: the second industrial revolution. Electric power transformed industry, increasing throughput rates, and mechanizing processes which had, until then, resisted inanimate power. Hand-held, electric power tools are a case in point. Electric drills, saws, and planners replaced hand drills, saws and planners.

Like the first, the second industrial revolution can be deconstructed into two sub-revolutions, the first occurring at the turn of the century and lasting up until WWII and the second staring in 1945 and ending in 1975. The first was characterized by the shift from either steam engine-driven or hydraulic-driven shafting and belting to electric motor drive, while the second was characterized by energy deepening in the form of multiple machine speed-ups.

This chapter proceeds in the same way as the last. We begin by examining the impact of electric power on Western industrialized economies, paying particular attention to the manufacturing sector. This is then followed by a look at its impact on political economy. How did the economics profession view this shock? More versatile than belting, gearing and shafting, electric drive de facto removed the energy constraint, setting the stage for what turned out to be 75 years of energy deepening.

Getting to the Garden of Eden, however, was not without setbacks. As we showed in Chapter 1, increasing quanta of power/energy are not a sufficient condition for growth. More specifically, a condition for output growth given the necessary productive capacity is commensurate money income growth. This raises the problem of underincome, specifically, the problem of money income growth in a producer-merchant exchange environment. As we argue in this Chapter, the shift from steam drive to electric drive and the ensuing increase in productive capacity, productivity, etcetera, resulted in an acute case of underincome. Productive capacity in U.S. manufacturing increased faster than money income in the 1910's, 1920's and 1930's. Drawing heavily from my earlier work (Beaudreau 1996a), we argue that the Smoot-Hawley Tariff Act of 1930, and the National Industrial Recovery Act of 1933 were policy measures aimed at resolving the problem of underincome in the U.S. economy in the 1920's and 1930's.

This then leads us to examine both the reaction to this new energy technology by political economists, and, secondly, its impact on political economy. As can be easily surmised, political economists in general were unable to identify the technological shock, let alone study its consequences. There were, however, exceptions, including University of Pennsylvania economics professor Simon Nelson Patten, Columbia University economics professor Rexford G. Tugwell, Columbia University economics professor Thornstein Veblen, British chemist and Nobel Prize laureate, Frederick Soddy, and Columbia University engineering professors Walter Rautenstrauch and Howard Scott. Finding turn-of-the century political economy to be largely irrelevant, owing to what they perceived of as ubiquitous abundance, not scarcity, they set out to replace the "economics of scarcity" by what Stuart Chase referred to as the "economics of abundance" (Chase 1933), based on energy. Man's ability to extract increasingly greater quanta of energy/force from his environment, they maintained, had transformed the problem of economics from one based on scarcity to one based on abundance. One could argue that political economy, being born of energy shocks (i.e. the steam engine), had never been about scarcity, but rather was, from its inception, about abundance. For example, Smith's Wealth of Nations can be viewed as a "how to" guide to abundance, one focusing, for the most part, on the "extent of the market" and specialization. That political economy is essentially British in origin is, in my view, a testimony to this. Nineteenth-century Britain had reached a level of material civilization unparalleled in history, a fact not lost on Alfred Marshall. It therefore follows that had economics been born of scarcity, then it most certainly would not have been British.

For the most part, these writers were ignored. Sitting pretty in its poorly-fitting, newly cut suit (i.e. Marshallian economics), neoclassical political economy failed to take heed. The second industrial revolution was ignored. However, its effects could not be. The ensuing breakdown of the Gold Standard, the Smoot-Hawley Tariff Act of 1930, the Great Depression, the National Recovery Administration, and the New Deal served to cast a long shadow of doubt on classical, radical, and neoclassical political economy as relevant "sciences of wealth." Lessons not learned from the past (i.e. the 19th century) had returned with Santayanian vengeance to haunt the western world.

The fall was great, but not fatal, as members of the scientific community, appalled and, in some cases, outraged by the irony of poverty in the face of such great potential, set out to rebuild the science of wealth. The Continental Committee on Technocracy, a collection of scientists and engineers, proposed a new science of wealth, founded on energy. The result was, for the most part, a refined version of Claude de Saint-Simon's 19th century blueprint of a society governed by scientists and engineers.

We begin by examining the impact of electric drive on output, distribution, and exchange using the analytical framework developed in Chapter 1. This will then be followed by a look at how these changes affected political economy in general. Whereas steam power revolutionized production processes by moving production out of "cottages" and into steam power-driven "factories," electric power revolutionized production processes in two fundamental ways. First, by replacing steam engines with electric motors, throughput rates were more easily controlled, and more importantly, increased. To the naive observer, nothing will have changed, yet, output may rise by 100 percent as the speed is doubled. Second, by making inanimate power available on a smaller scale, production processes which had, up until then, resisted the shift to steam power inanimate power succumbed. In the former case, the required capital actually decreased as cumbersome belting, gearing and shafting was replaced by electric motors.

Consequently, the introduction of electric drive went largely unnoticed in political economy, where, for the most part, technological change was associated with new, often times, more costly machinery. The effects of electric power on productivity, output, employment and expenditure, however, were not ignored. Last, we examine the indirect effects of the second industrial revolution on economic theory. We maintain that the fields of industrial organization, organizational behavior, industrial relations, Keynesian political economy (macroeconomics) are by-products of the second industrial revolution, more specifically of the problem of exchange (underincome), and the problem of distribution.

The First Second Industrial Revolution

Electricity Production in the Early 20th Century

As pointed out earlier, electric power is a misnomer, electricity not being a source of energy, but rather an enabling technology.[1] Early developments in electro-magnetics bear this out. When Michael Faraday discovered that an electric force exposed to a magnet repelled the latter, the force in question had been generated by the thermal energy in his body. Throughout the ensuing period, electric power continued to be generated from non-electric (i.e. non-atomic) sources. Only recently has Homo sapiens-sapiens learned how to tap the energy contained in the atom. Atomic power, however, represents but a fraction of current world total power consumption, the bulk coming from fossil-fuel and hydraulic-based energy sources, both of which are both based on solar radiation, and ultimately, the nuclear fusion of the sun's abundant supply of hydrogen into helium.

Table 3.1
Installed Generating Capacity, United States 1902 1970

Year	Total*	Utilities*	Industrial*	Steam*	Hydro*
1902	2,987	1,212	1,775	1,847	1,140
1907	6,809	2,709	4,100	4,903	1,906
1912	10,980	5,165	5,815	8,186	2,794
1917	15,494	8,994	6,500	11,608	3,886
1920	19,439	12,714	6,725	14,635	4,804
1921	20,605	13,519	7,086	15,603	5,002
1922	21,317	14,192	7,125	16,088	5,229
1923	23,235	15,643	7,592	17,553	5,682
1924	25,923	17,681	8,242	19,699	6,224
1925	30,087	21,472	8,937	22,937	7,150
1926	32,936	23,386	9,550	25,286	7,650
1927	34,574	26,647	9,495	26,647	7,927
1928	36,782	27,805	8,977	27,982	8,800
1929	38,708	29,839	8,869	29,783	8,925
1930	41,153	32,384	8,769	31,503	9,650
1931	42,287	33,698	8,589	32,097	10,190
1932	42,849	34,387	8,462	32,591	10,258
1933	43,037	34,587	8,450	32,707	10,330
1934	42,545	34,119	8,426	32,200	10,330
1935	42,828	34,436	8,392	32,429	10,399
1936	43,582	35,082	8,500	32,545	11,037
1937	44,370	35,620	8,750	33,184	11,186

1938	46,873	37,492	9,381	35,191	11,186
1939	49,438	38,863	10,575	35,932	12,075
1940	50,962	39,927	11,035	37,138	12,304
1941	53,995	42,405	11,590	39,474	12,912
1942	57,237	45,053	12,184	41,593	13,947
1943	60,539	47,951	12,588	43,840	14,991
1944	62,066	49,189	12,877	44,637	15,696
1945	62,868	49,189	12,877	45,248	15,892
1946	63,066	50,319	12,749	45,442	15,828
1947	65,151	52,322	12,829	47,242	15,956
1948	69,615	56,560	13,055	50,751	16,635
1949	76,570	63,100	13,470	56,472	17,662
1950	82,850	68,919	13,931	61,495	18,674
1951	90,127	75,775	14,352	67,372	19,870
1952	97,312	82,227	15,085	72,620	21,416
1953	107,354	91,502	15,852	80,960	23,054
1954	118,878	102,592	16,286	91,250	24,238
1955	130,895	114,472	16,423	101,698	25,742
1956	137,342	120,697	16,645	107,251	26,386
1957	146,221	129,123	17,098	114,660	27,761
1958	160,651	142,597	18,054	126,625	30,089
1959	175,000	157,347	17,653	139,073	31,884
1960	186,534	168,569	17,965	149,161	33,180
1961	199,216	181,312	17,904	158,588	36,301
1962	209,576	191,747	17,829	167,015	38,162
1963	228,757	210,549	18,208	183,348	40,928
1964	240,471	222,285	18,186	193,026	42,899
1965	254,519	236,126	18,393	205,423	44,490
1966	266,816	247,843	18,973	216,309	45,691
1967	288,185	269,252	18,933	234,195	45,691
1968	310,181	291,058	19,123	252,975	51,874
1969	332,606	313,349	19,257	273,534	53,447
1970	360,327	341,090	19,237	298,803	55,751

* '000 kW

Source: U.S. Department of Commerce (1975), S74-S77.

Table 3.1 provides data on the installed
generating capacity in U.S. electric utilities and
industrial generating plants from 1902 to 1970, by
type (i.e. hydro, steam and internal combustion).
In 1902, total installed generating capacity stood
at 2,987,000 kilowatts. By 1930, it had reached
41,828,000 kilowatts, a 1,277 percent increase,
which corresponds to an annual average annual
increase of 42 percent. By 1960, it stood at
186,534,000 kilowatts, which corresponds to a 345
percent increase over 1930. Broken down by type, we
find that in 1902, 38 percent of total installed
capacity was in the form of hydro, while the
remaining 62 percent was in the form of steam (i.e.
turning steam turbines). By 1960, hydro had lost
considerable ground to steam. Of the 186,534,000
kilowatts of total installed capacity in 1960, 18
percent (i.e. 33,180,000 kilowatts) was in the form
of hydro, while the remaining 82 percent was in the
form of steam (i.e. 143,116,000 kilowatts).[2]

As the numbers in Table 3.1 indicate, most of the growth in electricity-generating capacity in the United States in the 20th century in general and the early 20th century in particular, was coal-fossil fuel-based. This was also true of other industrialized nations such as the United Kingdom, Germany, and France. The point is that the second industrial revolution did not witness the emergence of a new power source, but rather, the emergence of a new power-drive technology, namely electricity.

Electric Drive

Few were the sectors of the U.S. economy not affected by the new power enabling technology. Production processes that hitherto had been driven by belting, gearing and shafting transmitting steam and hydraulic power, were converted to electric drive, resulting in substantially higher throughput rates. Those that had previously resisted "mechanization" succumbed. Among these were the mining sector, the petro-chemical sector, and the material handling sector in general. Electric motor-powered conveyor belts and pumps increased throughput rates in these sectors, resulting in vastly greater productivity. Analytically speaking, the introduction of electric drive should be viewed as an extension of the first industrial revolution where, as argued earlier, inanimate power replaced animate, muscular power. Now, however, the chemical energy contained in fossil fuels would drive electro-magnetic generators, the output of which would, in turn, drive production processes.

This section examines the effects of this far-reaching change on production, exchange, and distribution using the analytic framework developed in Chapter 1. As we shall argue, the shift from cumbersome belting, shafting and gearing-based drive to electric drive, while a watershed in the history of power drive technology (Nye 1990; Rosenberg 1983; Devine 1990), went largely unnoticed in political economy. Unlike the shift from muscular to steam power which, to most, defined the industrial revolution, it failed to generate much interest among political economists. Working with models in which capital and labor were the only factors of production, the shift to a new power drive technology was ignored, especially in light of the fact that, in many cases, electric-power drive was less capital intensive (Devine 1990).

By allowing for increased levels of energy consumption, the shift to electric drive increased output considerably. Few were the sectors which were not affected. Productivity per lower-level

supervisor (machine operatives) increased, in some
cases doubling. Once again, this raised the problem
of distribution. Who would appropriate the new set
of energy rents? Also, it raised the problem of
exchange, specifically, the problem of money income
growth. More productive firms have no private
incentives to increase wage income. Merchants, in
turn, have no private incentives to increase orders
from consumer and capital goods firms. Since
profits are a residual income form, it follows that
both wage and profit income fail to rise.

Electric Drive, Energy Consumption and Productivity

According to Warren D. Devine Jr., the shift from
shafting, belting and gearing drive to electric
drive stands as one of the most rapid and complete
transitions in "energy use" in history.

> Perhaps the most rapid and complete transition
> in energy use was the shift from steam power to
> electric power for driving machinery. Steam
> power prevailed at the turn of the century,
> with steam engines providing around 80 percent
> of the total capacity for driving machinery.
> By 1920, electricity had replaced steam as the
> major source of motive power, and by 1929—just
> forty-five years after their first use in a
> factory—electric motors provided 78 percent of
> all mechanical drive. (Devine 1990, 21)

While ignored by the economics profession, this
change galvanized the attention of the engineering
profession, broadly defined. Process and power
engineers could hardly contain their enthusiasm.
Consider, for example, the following quote taken
from a speech by Matthew S. Sloan, President of the
New York Edison Company, to the annual dinner of
trust companies in Chicago in February 1929:

> Mr. Sloan compared this age which he termed the
> "new industrial revolution" with "the
> industrial revolution" in the eighteenth
> century, when the steamboat and locomotive came
> into use. As steam brought in the machine era,
> electricity, he said, has brought in the era of
> mass production which has so greatly affected
> the general economic situation and social
> conditions. Thus, electricity, he said, is
> responsible for our present production. With
> all its attendant circumstances of lowered unit
> costs, lowered prices, increased wages,
> intensifying merchandising, wider markets, and
> a higher standard of living. Electricity-
> motivating machinery has multiplied the working

power of the nation many times, he said, and
the generating stations of the country now have
a capacity of 35,000,000 horsepower, of the
ability to do the work of about 350,000,000
men. In 1900, the generating capacity was only
3,000,000 horsepower. (New York Times, February
15, 1929)

In the same year, the President's Conference on
Unemployment, chaired by Herbert C. Hoover,
identified the electrification of U.S. industry as
the "single most important change in U.S. industry."
In its report, it described this far-reaching change
as follows:

> Characteristic also has been the rise in the
> use of power—three and three-quarters times
> faster than the growth of population—and the
> extent to which power has been made readily
> available not alone for driving tools of
> increased size and capacity, but for a
> convenient purposes in the the smallest
> business enterprise and on the farm and in the
> home. Factories no longer need to cluster
> about the source of power. Widespread
> interconnection between power plants, arising
> out of an increasing appreciation of the value
> of flexibility in power and made possible by
> technical advances during recent years, has
> created huge reservoirs of power so that
> abnormal conditions in one locality need not
> stop the wheels of industry. The increasing
> flexibility with which electricity can be
> delivered from power has enabled manufacturers
> and farmers to meet high labor costs by the
> application of power-driven specialized
> machines; and, power in this flexible form has
> penetrated into every section of the United
> States, including many rural areas. The survey
> shows that as a nation we use as much
> electricity as all of the rest of the world
> combined. Through the subdivision of power,
> the unskilled worker has become a skilled
> operator, multiplying his effectiveness with
> specialized automatic machinery and processes.
> (National Bureau of Economic Research, 1929,
> xi)

The shift to electric drive is easily formalized
in terms of E-O production analysis. Specifically,
electric drive, by lifting the constraints imposed
by belting, shafting and gearing transmission
technology, allowed for greater inanimate energy
consumption per period of time (i.e. $E(t)$). Higher
energy consumption per unit time period increased
throughput rates throughout all sectors of U.S., and
for that matter, in industries the world over,
increasing conventionally-measured productivity.

Output   per   lower-level   supervisor   (i.e.
conventionally-defined labor productivity) increased
substantially.   In sectors where water, wind, or
steam power had failed to penetrate, electric drive
provided a flexible, made-to-scale, source of power,
increasing productivity.   As pointed out earlier,
and highlighted in these quotes, there was hardly a
sector of the U.S. economy that had not been
affected in one way or another by electric drive.
     Consider, for example, the mining sector, which
had,   throughout   the   19th   century,   resisted
"mechanization."   Historian David E. Nye describes
the application of electricity in the mining sector
as follows:

> Mine owners had many uses for electricity.
> Electric lights gave safe illumination that did
> not exhaust scarce oxygen supplies.   Electric-
> alarm systems signaled danger or disaster.
> Electric drills were more portable than other
> drills: "Where electric power is used, small
> wires take the place of cumbersome pipes
> necessary for the transmission of steam or
> compressed air." Portable electric pumps often
> replaced steam engines, to keep pits free from
> water.   An   electric   hoist   had   similar
> advantages over a steam-driven hoist; it was
> "more easily installed, and when in place takes
> up much less room than a steam outfit of the
> same hoisting capacity. It does away with the
> boiler,   coal   bins,   and   piping."....   The
> greatest single change electricity made in the
> mines was the elimination of the mules, which
> were   replaced   by   squat,   powerful   electric
> locomotives.   With such equipment one company
> in Pennsylvania produced 11,000 tons of coal a
> month   in   the   1890's   without   using   a   single
> mule.   With   similar   equipment   New   York   and
> Scranton Coal Company saved five or six cents
> per ton of coal extracted and the Hillside Coal
> and Iron Company saved almost $20,000 per year
> on the cost of mules and laborers.".... At
> Green Ridge Colliery, for example, a station
> engineer, a motorman, and helper could run an
> electric   locomotive   that   replaced   six   mule
> drivers, four boy helpers, and seventeen mules.
> (Nye 1990, 205)

     No greater testimony to/of the prominent role of
electric drive in the rise of 20th century U.S.
industry is there than the "lamentations" of the
laggards at the time, the most important of which
was Great Britain.   As America's industrial might
increased with every additional megawatt of electric
power, Great Britain, mired in the glory of steam
and, more importantly, a web of regulation, could
only concede defeat. Beginning in 1916, a number of
committees were struck to study the problem.   The

first was the Coal Conservation Sub-Committee, set up in 1916 under Viscount Haldane, which, in turn, appointed a sub-committee to investigate the question of electric power supply in the U.K. Its report, issued in 1918, dealt in considerable detail with "the use of electric power in industry, and recommended the reorganization of the generation and main transmission on a regional basis under the central supervision of a "Board of Electricity Commissioners" with wide powers" (Self 1952, 35). In the same year, the Board of Trade appointed an Electrical Trades Committee to consider the position of the "electrical trades" after the war. Sir Henry Self summarized its findings as follows:

> Reference was made in strong terms to the crippling handicaps of the local and political considerations which had prevented Great Britain from reaping the fruits of the outstanding preeminence which it had received in original constructive research and development of electricity generation at the hands of pioneers such as Faraday, Wheatstone, Kelvin, Swan, Hopkinson, and many others. The loss of that outstanding lead, the history of industry in the intervening years, and the evidence taken during their examination of the position, led the Committee to the following conclusions:
> (i) That Government should recognize the dependence of the State, both from military and industrial standpoints, upon the supply of electrical energy as a "key industry."
> (ii) That the distribution of electrical energy should be regarded no longer as parochial but as a national question of urgent importance.
> (iii) that the present system of electrical generation and distribution is behind the times and is a serious handicap in international competition.
> (iv) That the present conditions are mainly due to faulty legislation and to divided and therefore weak executive control.
> (vii) That the determination of questions concerning concentration of generating plant with the resulting economy of coal and other savings requires immediate attention.
> (ix) That only by such steps can the electrical manufacturing industry of this country be fully developed, not only for the home trade, but as a consequence of the great industry now maturing overseas; that the gain to the State from a well-planned scheme of reconstruction will be inestimable; and that the items which are capable of reasonable calculation, such as saving in fuel, reduction in factory costs, and increased output will together represent not less than 100,000,000 pounds per annum (Self 1952, 35).

What is particularly noteworthy is the reference
made in point (ix) of the "great industry now
maturing overseas." This clearly demonstrates that
the Electrical Trades Committee was well aware, as
early as 1916, of the threat posed by the
electrification of U.S. industry. Similar
conclusions were reached by the Williamson
Committee, struck by the Board of Trade in 1917:

> (1) That when British industry is subjected to
> the test of keen international competition
> after the war, its success will depend upon the
> adoption of the most efficient methods and
> machinery, so as to reduce manufacturing costs
> as much as possible.
> (2) That a highly important element in reducing
> manufacturing costs will be the general
> extension of the use of electric power supplied
> at the lowest possible price, and it is by
> largely increasing the amount of power used in
> the industry that the average output per head.
> and, as a consequence, the wages of the worker
> can be raised.
> (5) That a comprehensive system for the
> generation of electricity, and, where
> necessary, reorganizing its supply should be
> established as soon as possible (Self 1952,
> 37).

In 1924, David Lloyd George, then Liberal member
of Parliament, struck an informal committee, the
purpose of which was to address the "inter-linked
questions of coal and electricity." The results of
the inquiry were published in a report entitled
"Coal and Power Inquiry," which called for
compulsory powers of acquisition, coordination and
regulation to enable the "Electricity Commissioners"
to grant to approved bodies the right of supplying
power within substantial defined areas."(Self 1952,
52). H.H. Ballin summarized the Inquiry's findings
as follows:

> With our various competitors going ahead
> swiftly in the direction of the greater
> utilization of power, a policy of preventing
> power development in industry would leave our
> workers in the position of having to compete on
> unequal terms so that the incompetence of
> management would have to be made up by the toil
> of the workers. It ushers in a vista of
> endless strikes, industrial trouble and
> internal strain. The way out is to be found in
> the direction of scientific production and
> utilisation of power (Lloyd George 1924b, 99).

How did the resulting energy deepening affect
conventionally-defined productivity in U.S.

manufacturing? That is, how did it affect $W(t)/S_1(t)$
(see Table 1.3)? As it turns out, this is a
particularly difficult question, given data
problems. The data for productivity in this period
are rather sparse and incomplete. This
notwithstanding, we proceeded in two distinct ways.
First, we considered the case of the Ford Motor
Company which, in 1913, moved from static assembly
to electric power-driven dynamic assembly, commonly
known as mass production. Second, various E-OPA
productivity measures developed in Chapter 1 were
estimated for this period.

The choice of the Ford Motor Company (FMC) as a
case study was based on its early use of electric
drive (Nye 1990; Beaudreau 1996a). It is important
to note that its founder, Henry Ford, had been chief
mechanical engineer at the Detroit Edison
Illuminating Company from 1886 to 1899, a period of
14 years. In earlier work (Beaudreau 1996a), we
argued that this period in Ford's life is key to
understanding of the development of high-throughput
continuous-flow manufacturing at the FMC and,
ultimately, in the United States. From then on,
Ford was, to put it mildly, a power zealot, as
evidenced by his lifelong friendship with Thomas
Edison, and his attempt at securing the rights to
the Muscle Shoals hydroelectric project in 1919.
His single most important contribution to production
technology, we argued in Beaudreau (1996a), was the
application of electric drive to production and
assembly processes, resulting in record increases in
productivity (Nye 1990; Beaudreau 1996a).

Table 3.2
The Ford Motor Company, Assembly Process Productivity

Process	Phase	Assembly Time	Index
Magneto	Initial	20m,0s	100
	I	13m,10s	66
	II	7m,0s	35
	III	5m,0s	35
Transmission Cover	Initial	18m	100
	I	9m,0s	50
Chassis Assembly	Initial	12h,28m	100
	I	6h,0m	48
	II	1h,33m	12

Source: Beaudreau (1996a),6.

To see this, consider Table 3.2, taken from
Beaudreau (1996a), which presents throughput data
measured in terms of time. Prior to the

introduction of electric drive-powered assembly
lines, it took, on average, 20 minutes to assemble a
magneto. In the initial phase of electric drive-
powered assembly lines (i.e. conveyor belts), this
was reduced to 13 minutes, followed by a further
reduction to 7 minutes, and ultimately, to 5
minutes. Productivity, it therefore follows, had
quadrupled. Even greater productivity gains were
achieved in chassis assembly where the required time
went from 12 hours, 28 minutes, to 6 hours, and
eventually, to 1 hour, 33 minutes.

Table 3.3
The Ford Motor Company, Production Data 1903-1916

Year	Employees	Model T's	Model T's/Empl.
1903	125		
1904	300		
1905	300		
1906	700		
1907	575		
1908	450		
1909	1,655	13,840	8.36
1910	2,773	20,727	7.47
1911	3,976	53,488	13.45
1912	6,867	82,388	11.99
1913	14,366	189,088	13.16
1914	12,880	230,788	17.91
1915	18,892	394,788	20.89
1916	132,702	585,388	17,90

Source: Beaudreau (1996a),152.

From a level of 8.36 in 1912, labor productivity
climbed, mercurially, to 20.89 in 1915, a 150
percent increase. Electric drive had more than
doubled productivity. In the 1926 edition of the
Encyclopaedia Britannica, Ford described "mass
production" in the following terms:

> Mass production is not merely quantity
> production, for this may be had with none of
> the requisites of mass production. Nor is it
> merely machine production, which also may exist
> without any resemblance to mass production.
> Mass production is the focusing upon a
> manufacturing project of the principles of
> power, accuracy, economy, system, continuity,
> and speed. (Ford 1926, 821)

The operant words here are power and speed. Mass
production, Ford style, was about speed.[3] In little
time, electric drive as applied at the Ford Motor
Company spread throughout U.S., and, for that

matter, the world over.   According to historian
David S. Hounshell:

> The story of mass production at the Ford Motor
> Company was not something only historians of a
> later generation would delve into and try to
> understand. Henry Ford's contemporaries, many
> of whom were competitors, closely watched the
> doings at Highland Park, attempting to
> understand and emulate the revolutionary
> developments.   Henry Ford encouraged their
> interest.   Unlike the Singer Manufacturing
> Company, the Ford Company was completely open
> about its organizational structure, its sales,
> and its production methods....   As a
> consequence of Ford's openness, Ford production
> technology diffused rapidly throughout American
> manufacturing. (Hounshell 1984, 260)

Consumption of electric power in U.S.
manufacturing increased dramatically in the ensuing
years. Referring to Table 3.4, we see that from a
level of 9,250 million kilowatt hours in 1912, it
had more than doubled by 1917 (i.e. 20,750 million
kilowatt hours), and, doubled again by 1927. Cast
in per-worker terms, in 1912, each manufacturing
employee supervised 1,111 kilowatt hours of electric
power. This doubled by 1917, tripled by 1923, and
quadrupled by 1926.[4]   Conventionally-defined
productivity increased in-step.   Productivity
indexes $(W(t)/S_1(t))$ for this period show a marked
increase. For example, the NBER index goes from a
level of 29.2 in 1912. to a level of 46.5 in 1926, a
60 percent increase.

Table 3.4
U.S. Manufacturing Data 1912-1945

Year	Electric Power	Employees	Ratio	NBER
1912	9,250	8,322	1,111	29.2
1917	20,750	9,872	2,101	31.7
1920	26,913	10,702	2,514	32.0
1921	23,993	8,262	2,904	36.8
1922	27,364	9,129	2,997	41.8
1923	32,585	10,317	3,158	40.2
1924	34,967	9,675	3,614	42.8
1925	39,725	9,942	3,995	45.6
1926	46,350	10,156	4,563	46.5
1927	51,012	9,996	5,103	47.6
1928	52,699	9,942	5,300	49.7
1929	55,122	10,702	5,150	52.0
1930	53,930	9,562	5,640	52.3
1931	50,410	8,170	6,170	54.0
1932	43,504	6,931	6,276	50.5
1933	46,561	7,397	6,294	54.9
1934	50,593	8,501	5,951	57.4
1935	56,706	9,069	6,252	61.2

1936	62,949	9,827	6,405	61.6
1937	64,757	10,794	5,999	60.7
1938	58,452	9,440	6,191	59.9
1939	70,518	10,278	6,861	65.4
1940	83,276	10,985	7,580	68.7
1941	104,037	13,192	7,886	71.2
1942	122,762	15,280	8,034	72.4
1943	143,995	17,602	8,180	73.4
1944	145,015	17,328	8,368	72.5
1945	134,955	15,524	8,693	71.5

*000,000 kilowatt hours.
** '000.
***NBER
Source: U.S. Department of Commerce (1975), Series D130, S124, D802.

Like steam power a century and a half earlier, the shift to electric drive increased throughput rates, and, consequently, conventionally-defined labor productivity. It is important to point out, however, that conventionally-defined labor was not more productive; rather, it was being called upon to supervise the more and more work done, in this case, by inanimate energy. The second industrial revolution was, as such, a continuation of the first industrial revolution. Like the first, the key ingredient was power.

Electric Drive, Energy Deepening and Early 20th Century Political Economy

As argued in the previous chapter, 19th-century political economy was the result of energy innovations, namely water power and "fire power" in the form of water-powered and steam power-driven machinery. Classical, radical, and neoclassical political economy, it therefore follows, were an outgrowth of the "machine age," dating back to the late 18th century. The irony, as pointed out earlier, is the absence of energy per se in formal models of production. Energy innovations were the defining changes; yet, energy was nowhere to be found. This, we argue, is crucial to understanding the role of electric drive and energy deepening in early 20th century political economy.

Theoretical models of production (e.g. Cobb-Douglas) in this period focused exclusively on capital and labor (Cobb and Douglas 1928). As a result, if energy in general and electric power in particular were to make their way into the science of wealth, it would have to be through the back door, metaphorically speaking. In this section, we show that the shift to electric drive and the ensuing energy deepening provoked an important chasm

in political economy, one based on energy.
Mainstream political economy, based largely on the
work of Alfred Marshall, William Stanley Jevons, and
Leon Walras, was, for the most part, unable to
identify the technology shock. We refer to this body
of literature as the "economics of scarcity,"
scarcity being the defining problem. By contrast,
in the 1920's and 1930's, there emerged a group of
political economists for whom the central economic
problem was not scarcity, but rather abundance.
Best associated with the names of Simon Nelson
Patten, Thornstein Veblen, Rexford Tugwell, F.G.
Tyron, Stuart Chase, and Edward A. Filene, to
mention a few, this group argued that technology,
especially developments in electric power
generation, distribution and consumption, had
increased America's ability to produce, far beyond
imagination. The problem, as they saw it, rested
with its ability to exploit this new, higher
ability. In short, America had failed to exploit
its new higher capacity. While the reasons for this
differ among writers, they share one fundamental
belief, namely that the central economic problem was
no longer scarcity, but abundance. We refer to the
resulting body of literature as the "economics of
abundance," described by Stuart Chase's in his 1935
book entitled The Economics of Abundance as follows:

> Two men are lost on a great desert. One has a
> full bottle of water, the other a bottle
> quarter filled. As they move wearily onward,
> hoping for an oasis, justice demands that they
> pool the water supply and share it equally.
> Failure to do so will undoubtedly result in a
> fight. Now let us transport these two men to a
> row boat on Lake Superior. Again, they are
> lost, and again, one has a full bottle of
> water, and one a bottle quarter full. The full
> bottle man refuses to share and a battle
> ensues. Maniacs! There is plenty of fresh water
> over the side of the boat.
> The Desert is the Economy of Scarcity; the
> lake, the Economy of Abundance. The choice
> between sharing or fighting is chronic in the
> former, pointless in the latter. Today,
> throughout western civilization, men in boats
> are fighting, or preparing to fight, for fresh
> water. They do not know they are in boats;
> they think they are still on camels. The lake,
> as we have seen in the previous chapter, is not
> limitless, but nobody need go thirsty. (Chase
> 1933, 51)

Scarcity versus Abundance

The prominent role of energy as the key element in the economics of abundance, as defined by Patten, Tugwell, Tyron, Rautenstrauch, Scott, and numerous others, and, secondly, the absence of energy in the corpus of classical, radical, and neoclassical political economy, raise a number of interesting questions. For example, why is it that classical (radical and neoclassical) political economy is about scarcity when "fire power," like electric power in the 20th century, had lifted and pushed back the energy constraint (i.e. animate, muscular energy)? After all, Great Britain in the 19th century was the richest country in the Western world; the U.K. worker the best paid. Yet, ironically, the country in which wealth abounded, relatively speaking, yielded a science of wealth based on scarcity.

Part of the answer, we believe, lies in (i) 19th-century developments in U.K. agriculture, and (ii) the U.K.'s growing dependency on foreign nations for its industrial feedstocks (cotton, silk). While abundance reigned in U.K. industry (manufacturing value added), scarcity reigned in early 19th-century U.K. agriculture. Population growth in the late 18th century and throughout the 19th century had increased the demand for food, increasing price, and, consequently, putting upward pressure on real wages (Ricardo, Malthus). Higher wages, whether actual or anticipated, implied lower profits, and, ultimately, lower savings, investment, and growth. The United Kingdom's ability to grow, it therefore follows, was intricately linked to its ability to feed itself. Abundance and scarcity coexisted throughout the early-to-mid 19th century in the U.K., the latter coming to dominate political economy.

Another factor is the U.K.'s almost complete dependence on foreign nations for its industrial feedstocks (cotton, wool, silk, linen, etcetera). The faster the U.K. economy grew, the more likely was it to come up against the proverbial upward-sloping Ricardian rent curves. In short, its almost total dependence on agriculture (for food and industrial feedstocks) provides another reason for the triumph of "scarcity" over "abundance."

As it turned out, the second industrial revolution occurred in a food-abundant country, namely the United States of America.[5] Unlike the United Kingdom where the supply of arable land was limited, the United States had surplus land. Further lessening the food problem in the early 20th century was the coming of age of the internal combustion engine which replaced horses, mules and oxen as a source of motive power, thus leading to conditions of excess supply in agricultural markets (i.e. throughout the 1920's). Electric power and the internal combustion engine, it therefore follows, combined to transform the economic problem in the U.S. from one of scarcity to one of abundance.

The discussion in the next section will be organized these two approaches to economics in the early 20th century, namely, the Economics of Scarcity and the Economics of Abundance. The former is epitomized by Marshallian price theory, which, at the time, dominated political economy in the United States. As will soon become apparent, rarely did the "twain" meet.

Electric Drive, Energy Deepening and Production

The Economics of Scarcity

By definition, the economics of scarcity refers to classical and neoclassical political economy. As shown in the previous chapter, classical and neoclassical production theory focuses, for the most part, on two factor inputs, namely capital and labor. Output is increasing in capital and labor; energy is ignored altogether. It being the case that early 20th century U.S. political economy was predominantly Marshallian in origin, it comes as no surprise to find energy in general and electric power in particular to be absent from mainstream economics. It being the case that electric power was less capital intensive than steam power, the shift to electric drive witnessed a decrease in investment, and, hence, in the measured level of capital (Devine 1990). Hence, to the orthodox, mainstream U.S. political economist, if anything, there was reason to believe that output would decrease, not increase.

This is not to say, however, that all orthodox mainstream political economists were oblivious to the power revolution. While most failed to seize the importance of electric power in industry, especially in manufacturing, there were those, who, owing either to a keen sense of observation, or to the nature of their job, seized its importance. The problem, however, was the lack of an appropriate analytical framework. As is the case today in growth accountancy (Denison 1962,1985) where non-traditional factors (e.g. education, infrastructure, trade) are put in the residual, in the 1920's, electric power was considered a residual cause.

This is the case of Columbia University professor Rexford G. Tugwell and Yale University professor Irving Fisher. In a book entitled Industry's Coming

of Age, published in 1927, Tugwell listed a number
of changes to industry, including "the bringing into
use of new and better power resources more suited to
our technique, more flexible and less wasteful."

> The electrification of industry has now
> progressed to the extent of between 55 and 60
> per cent completion. So widespread an adoption
> of this new flexible means of moving things
> cannot have taken place without numerous
> secondary results in lowering costs,
> improvements in quality, and a heightened moral
> among workers. For the new power is not only
> cheaper to use; it is also cleaner, more silent
> and handier. On the whole, the electrification
> of industry must be set down as the greatest,
> single cause of the industrial revolution.
> (Tugwell 1927, 182)

In The Stock Market Crash and After, Irving
Fisher ranked the electrification of U.S. industry
as one of the contributing factors behind the stock
market's rise in the late 1920's. Changing economic
fundamentals, he maintained, were the underlying
cause.

> But after 1919, something happened. The
> implications of which are not yet sufficiently
> gauged. It was of enough significance to cause
> President Hoover's Committee on Recent Economic
> Changes to remark that "acceleration rather
> than structural change is the key to an
> understanding of our recent economic
> developments." The committee added: "But the
> breath of the tempo of recent developments
> gives them new importance." What happened was
> indicated by the fact that in the United
> States, eight million, three hundred thousand
> workers produced in 1925 one quarter more than
> nine million wage workers turned out in 1919.
> The new indexes of the Federal Reserve Board
> measuring production record this gratifying
> advance which reflects an increase in the
> American standard of living. The indexes
> cover, directly and indirectly, four-fifths of
> the industrial productivity of the nation
> directly in about thirty-five industries, and
> collaterally, in many more. The general volume
> of production had increased between 1919 and
> 1927 by 46.5 percent, primary power by 22
> percent, and primary power by wage earner by
> 30.9 percent (between 1919 and 1925) and
> productivity per wage worker by 53.5 percent
> between 1919 and 1927. (Fisher 1930, 120)

What is important to note is the nature of the
handful of references to electric drive and
electricity, namely as orthogonal to production

theory.   Not having a general theory of production
(i.e. such as that developed in Chapter 1),
references to electric drive and electricity were
often included in long, exhaustive lists of non-
orthodox causes of productivity growth.   For
example, Fisher lists electric drive among a number
of causes, including "less unstable money, new
mergers, new scientific management, and the new
labor policy of waste saving" (Fisher 1930, 129).

There were also those who owing to the nature of
their work (job task) were cognizant of the power
revolution.   This is the case of F.G. Tyron of the
Institute of Economics (Brookings Institution), who
in 1927 lamented the little attention paid to power
by economists, and Woodlief Thomas of the Federal
Reserve Board, who attributed the "Increased
Efficiency of American Industry" to the increasing
use of machinery and power.   According to Tyron:

> Anything as important in industrial life as
> power deserves more attention than it has
> received by economists.   The industrial
> position of a nation may be gauged by its use
> of power.   The great advance in material
> standard of life in the last century was made
> possible by an enormous increase in the
> consumption of energy, and the prospect of
> repeating the achievement in the next century
> turns perhaps more than anything else on making
> energy cheaper and more abundant.   A theory of
> production that will really explain how wealth
> is produced must analyze the contribution of
> this element of energy.
>
> These considerations have prompted the
> Institute of Economics to undertake a
> reconnaissance in the field of power as a
> factor of production.   One of the first
> problems uncovered has been the need of a long-
> time index of power, comparable with the
> indices of employment, of the volume of
> production and trade, of monetary phenomena,
> that will trade the growth of the factor of
> power in our national development. (Tyron 1927,
> 281)

One year later (i.e. in 1928), Woodlief Thomas of
the Division of Research and Statistics of the
Federal Reserve Board, published an article in the
American Economic Review entitled "The Economic
Significance of the Increased Efficiency of American
Industry," in which he attributes the "striking
changes" in American industry" to power-related
developments:

> Large-scale production is dependent upon the
> machine process, and the increasing use of
> machinery and power and labor-saving devices

> has accompanied the growth in size of
> productive units. The growing use of power in
> manufacturing, for example, is reflected in the
> increase in horsepower of installed prime
> movers. This does not tell the whole story,
> moreover, for owing to increased use of
> electricity, the type of power used is now more
> efficient—requiring less fuel and labor for
> its production. Out of a total installed
> horsepower in factories of thirty-six million
> in 1925, twenty-six million or 72 per cent was
> transmitted to machines by means of electric
> motors, as compared with 55 percent in 1919, 30
> per cent in 1909, and only 2 per cent in 1899.
> Between 1899 and 1925 horsepower per person
> employed in factories increased by 90 percent
> and horsepower per unit of product increased by
> 30 percent.... Power has been substituted for
> labor not only through machines of production
> but also in he form of automatic conveying and
> loading devices. (Thomas 1928, 130)

Unfortunately, Tyron and Woodlief were the
exceptions. Mainstream political economists were,
for the most part, unaware of the profound changes
thrust upon U.S. industry by the application of a
new power form, electricity. This oversight, we
maintain, is what prompted a number of young
political economists, scientists and engineers to
reject orthodoxy in favor of a new science of
wealth, one born of energy and based on abundance.

The Economics of Abundance

Ironically, the study of political economy, born
of energy innovations of the late 18th century, was
unable, a century later, to identify, let alone
analyze, subsequent energy innovations. Engineers
and accountants, businessmen and politicians
revelled in the new power source; political
economists, on the other hand, were oblivious to it.
This, we argue, is what prompted, in the 1920's and
especially 1930's, to a veritable deluge of
contributions in the economics of abundance. Among
the leading authors—and there were many—were Edward
A. Filene, Stuart Chase, Henry Ford, John Atkinson
Hobson, Howard Scott, Frederick Soddy, Rexford
Tugwell, and Thornstein Veblen. Defining the
economics of abundance were (1), the role of energy
in general, and electric power in particular in
production processes, and consequently in early 20th
century U.S. economic growth, (2) the purported
flawed nature of the for-profit economy,
distribution system, and/or money supply process,
and (3) the call for change, whether it be in the
conduct of business or government regulation.

Regarding the role of energy, nearly every treatise begins by extolling the virtues of the new "power age." Consider, for example, the following excerpt from Howard Scott's Introduction to Technocracy, published in 1933:

> A century ago these United States had a population of approximately 12,000,000 whereas to-day our census figures a total of 122,000,000—a tenfold increase in the century. One hundred years ago, in these United States, we consumed less than 75 trillion British thermal units of extraneous energy per annum, whereas in 1929, we consumed approximately 27,000 trillion British thermal units—an increase of 353 fold in the century. Our energy consumption now exceeds 150,000 kilogram calories per capita per day; whereas in the year 1800 our consumption of extraneous energy was not less than 1600 or more than 2000 kilogram calories per day.... The United States of our forefathers, with 12,000,000 inhabitants, performed the necessary work in almost entire dependence upon the human engine, which, as its chief means of energy conversion, was aided and abetted only by domestic animals and a few water wheels. The United States to-day has over one billion installed horsepower. In 1929, these engines of energy conversion, though operated only to partial capacity, nevertheless had an output that represented approximately 50 percent of the total work of the world. When one realizes that the technologist had succeeded to such an extent that he is to-day capable of building and operating engines of energy conversion that have nine million times the output capacity of the average single human being working an eight hour day, one begins to understand the acceleration, beginning with man as the chief engineer of energy conversion and culminating with these huge extensions of his original one-tenth of a horse power. Then add the fact that of this 9,000,000 fold acceleration, 8,766,000 has occurred since the year 1900. (Scott 1933, 42)

Another good example is the following excerpt taken from Chapter 1 of Stuart Chase's Economics of Abundance, entitled "Forty to One":

> Suppose that the thirteen million people living in the United States in 1830 had awakened on the morning of January 1, 1831, with forty times the physical energy they had gone to bed with the night before. An active picture meets the mind's eye; a very active picture. A lumberman can fell forty times as many trees in a week, a housewife sweep forty times as many

square feet of floor; forty barns can be built
in the time hitherto required for one—and
forty chests and forty chairs.   Porters can
transport forty times their accustomed load in
a day; weavers ply their shuttles forty times
as fast—if the shuttles can brook the strain;
and children raise forty times their normal
rumpus.

Assuming no increase in the invention of labor-
saving devices—and where would be the point with
such an exuberance of labor available—what might we
logically expect in the way of economic changes in a
culture essentially handicraft?   From an economy of
scarcity, with barely enough to go around, the young
republic would almost immediately enter an economy
of abundance. (Chase 1933, 1)
    In the same chapter, Chase defines "Abundance" in
terms of a series of propositions.

1. A condition where the bulk of the economic
work is performed not by men, but by inanimate
energy, drawn from coal, oil and water power.
Such a condition was reached in the United
States towards the close of the nineteenth
century, circa 1880.
2. A point at which living standards per capita
reach an average which is, at least
potentially, twice as high as ever obtained
under scarcity conditions. Reached circa 1900.
3. A point at which the curve of invention,
following, as it does, a geometric increase,
becomes the dominant factor in economic
life—precisely as the Nile was the dominating
factor in the economic life of Egypt.   Circa
1870.
4. A point at which the scientific method
supersedes the use and want of the craftsman in
the production of most material goods.   Circa
1900.
5. A point where output per man hour becomes so
great that total productive labor must
thereafter decline, even as output grows.   A
point at which labor ceases to be a measure of
output—as it always has been in preceding
ages. Circa 1920.
6. A point at which overproduction carries a
more serious threat to the financial system
than shortage. Circa 1880.
7. A point at which specialization has
destroyed all practicable local self-
sufficiency and made economic insecurity for
all classes latent, growing, and ultimately
intolerable, given no change in financial
methods. Circa 1900, with the closing of the
American frontier.
8. A point at which consumption becomes a
greater problem than production. Circa, 1920.
"Our economy," says F.L. Ackerman, "is so set

up that it produces goods at a higher rate that
it produces income with which to purchase
them."
9. A point at which the industrial plant is,
substantially, constructed, requiring
relatively smaller outlays for capital goods in
the future, and where pecuniary savings are not
only unnecessary in their old volume, but
seriously embarrassing, Circa, 1925.
10. A point where, due to the presence of the
technical arts, costs, prices, interest rates,
debts, begin a descent with zero as their
objective. Circa 1920. (Chase 1933, 12)

Unfortunately, these ideas were never formalized.
Scott, Chase, Tugwell, and others failed to move
beyond mere words in their quest for a new theory of
production. What, for example, was the role of
capital? What was the role of labor? Was each
productive in the physical sense? Was output a
function of energy only? Unfortunately, these
questions were never addressed, at least
satisfactorily, having been usurped in the public's
eye by the New Deal with its Keynesian-style
spending policies. The "Economics of Abundance,"
despite holding great promise scientifically and
politically, was virtually forgotten.

Indirect Effects on Economic Theory

The shift to electric drive and the ensuing
energy deepening in the form of higher electricity
consumption, like the shift from the domestic system
to the factory system some 150 years earlier, failed
to change the way political economists in general
(classical, neoclassical and radical) modeled
production. Capital and labor continued to be the
two sole factor inputs. Nonetheless, these
developments did leave their mark on political
economy, specifically on the study of industrial
organization, macroeconomics, and industrial
relations were the result, in large measure, of the
introduction of high-throughput, continuous-flow
production techniques. Here, we provide a brief
account of the role of electric drive and the
ensuing energy deepening in the development of the
field of industrial organization.

Industrial Organization

It is generally held that the study of industrial
organization begins, in earnest, with Edward
Chamberlin and Joan Robinson's writings on
monopolistic competition in the early 1930's.
Before, markets were classified either as

competitive or monopolistic/duopolistic. The relevant question, from a historical point of view, is what factors prompted this upheaval in price theory? Why was perfect competition, the staple of Marshallian price theory, out, and monopolistic competition, a combination of competition and monopoly, in? The answer to this question, as we shall attempt to demonstrate, is complex, having roots in Marshallian price theory, and in the electrification of U.S., and indeed, U.K. manufacturing.

Interest in price theory in the early 20th century was a decidedly British phenomenon. From J. H. Clapham's "On Empty Economic Boxes," which appeared in the September 1922 number of the Economic Journal to Piero Sraffa's "The Laws of Returns under Competitive Conditions" to Joan Robinson's "The Theory of Monopolistic Competition," what stands out is its uniquely British flavor. This only serves to beg the question why? Why was interest in price theory, specifically in the failings of Marshallian price theory, concentrated in the U.K.? The main reason, we argue, is the debate over the gold standard, and the apparent downward inflexibility of prices. World War we saw prices increase in the United Kingdom by roughly 200 percent, and by 100 percent in the U.S. A return to the Gold Standard at pre-war parity (i.e. $4.86 per pound) would require that prices in the U.K. fall substantially. From 1918 to 1922, they had not, forcing a debate over the workings of markets, and more importantly, over the relevance of Marshallian price theory. World War we price developments (i.e. increases, and decreases), as it turns out, constituted the first true test of Marshallian price theory.

Assuming that markets are organized competitively (perfect competition), a decrease in demand should, in the presence of an upward-sloping supply curve, result in a lower price. Should this not be the case, then either the model is flawed, or, at the very least, is incomplete. For example, if firms face increasing returns or decreasing costs, then a decrease in demand is less likely to lead to lower prices given what are higher unit costs. This explains, among other things, the plethora of work on returns to scale and increasing and decreasing cost industries in the 1920's.

In hindsight, one could argue that "the chickens had come home to roost." Marshallian price theory, being politically motivated, had ignored, for the most part, the reality which was high throughput, continuous-flow production processes, synonymous with the intensive use of inanimate energy. While upward-sloping supply curves may well be consistent with Neolithic, artisanal production processes, they were orthogonal to the large-scale textile mills of Northern England.   Marshall's desire to move away from a substance theory of value—and ultimately, the Marxist critique—, one could argue, had exacted a toll.   The events of the post-WWI period more than confirmed this.   Specifically, they set into motion a process which culminated in the study of industrial organization as an important branch of microeconomic theory.

While the original impetus for a re-examination of price theory, specifically of the role of the organization of industry, came from the U.K., the study of industrial organization is, historically speaking, a U.S. phenomenon.   Notwithstanding the pioneering contributions of Joan Robinson, Piero Sraffa, and Arthur Pigou, the defining contributions are American.   Among these are Edward Chamberlin's work on monopolistic competition, Edward Mason's work on industry structure and performance, Paul Sweezy's work on price rigidity, and Joe Bain's Structure-Behavior-Performance framework.   Why was this the case?   Why did U.S. political economists come to dominate industrial organization, and ultimately, price theory?   The answer, we maintain, is intimately tied to domestic developments, namely, the development of extremely high-throughput, continuous-flow, mass production techniques.   As pointed out earlier, the second industrial revolution was, in many regards, a continuation of the first.   Specifically, industries in which production processes were driven by steam saw throughput rates increase substantially with electric drive.   Industries which had, up until then, resisted the power revolution, succumbed, resulting in important productivity gains.   The assembly line is a case in point.

By the 1930's, U.S. industry had, for the most part, become monolithic, being characterized by "large-scale enterprise" (Sobel 1972).   Hence, it stands to reason that the need for a theory of pricing behavior by large-scale enterprise would have been more pressing in the U.S. than in the U.K.   One of the first to address this issue was John Maurice Clark, son of John Bates Clark, in a book entitled The Economics of Overhead Costs, published in 1923.   The following passage, the first paragraph of the Preface, sets the tone:

> This volume is a bit of research into the principles of dynamic economics.   It is an experiment in a type of economic theory which is largely inductive, which comes to grips with the dynamic movements and the resistances to movement, and the organized interrelations of parts, which make our economic world a dynamic

> social organism, rather than a static mechanism
> with an endless uniformity of perpetual motion.
> It studies the discrepancies between supply and
> demand; indeed the whole subject of the book
> might be defined as the study of discrepancies
> between an ever-fluctuating demand and a
> relatively inelastic fund of productive
> capacity, resulting in wastes of partial
> idleness and many other economic disturbances.
> Unused capacity is its central theme. (Clark
> 1923, ix)

With the advent of high-throughput, continuous
flow mass production techniques, new capacity
additions (whether add-ons or greenfield) were
increasingly lumpy. Capacity utilization rates, it
therefore follows, would have decreased as firms
found themselves with significant excess capacity,
at least in the short run. In time, demand growth
would increase capacity utilization rates, until the
while process began anew.

The problem raised by Clark is as real today as
it was then. How did large-scale firms price their
product in what is, essentially, a dynamic
environment. Price theory, for the most part, is
static in nature, owing, in large part, to
Marshall's "Principles." The fact of the matter,
however, is that with large-scale firms, costs (i.e.
average and marginal) are dynamic, fluctuating as
demand grows over time, capacity increases occurring
at discrete points in time.

What is particularly noteworthy of The Economics
of Overhead Costs is its timing, namely 1923, two
years after the depression of 1921, but six years
before the Great Depression. The bulk of work on
prices and costs in the U.S. dates to the 1930's,
and, should as such, be seen as a response to the
price inertia observed during the Great Depression.
Another noteworthy feature is its scope. The
Economics of Overhead Costs was more than a simple
treatise on overhead costs; it was, for all intents
and purposes, Clark's attempt at a "general theory
of economics" based on high throughput, continuous-
flow mass production. Table 3.6 presents its table
of contents, consisting of twenty-four chapters.
The topics range from "How and Why large Plants
Bring Economy," to "Labor as an Overhead Cost," to
"Overhead Costs and the Laws of Value and
Distribution." Economics of Overhead Costs should,
as such, be seen as an attempt, visionary at the
time, to provide an empirically-consistent theory of
high-throughput mass production and pricing.

Table 3.6
The Economics of Overhead Costs-Table of Contents

Source: Clark (1923),xi.

The Fall of the Economics of Abundance

As we have shown, the early 20th century witnessed the emergence of an alternative science of wealth, one based on abundance and not on scarcity. For reasons described above, the for-profit economy, as Veblen put it, had failed to translate potential output into actual output. The U.S. economy in the 1920's and 1930's, they maintained, was, paradoxically, suffering from abundance, not scarcity.

Unfortunately, this highly promising, truly scientific line of inquiry failed to make inroads into mainstream political economy, as evidenced by its total absence today. In fact, if anything, it

has become an object of scorn in the literature, as
evidenced by work bearing titles like "Cranks,
Heretics, and Macroeconomics in the 1930s." This
raises a number of questions, one of which being,
why? Why did the Economics of Abundance fail to
make inroads into mainstream political economy? Why
did a theory of production based on energy, the
cornerstone of classical mechanics and
thermodynamics, fail to make inroads into mainstream
production theory? These questions merit more
attention than we can accord them here.

The reasons, we believe, are many. In no
particular order, they are (1) its emphasis on
energy, which, at the time, was foreign, not to say
orthogonal, to production theory, (2) its total
eschewal of mainstream political economy, as best
illustrated by the writings of Thornstein Veblen and
the Technocrats, (3) its diagnoses of the Great
Depression, (4), its policy prescriptions, and (5)
the rise of Keynesian economics.

Perhaps the most striking feature of "Economics
of Abundance" is the role accorded to energy.
Energy was the starting point of production theory
in this literature. This, as it turns out, was
orthogonal to mainstream political economy, where
the emphasis continued to be on capital and labor.
Clearly, a veritable "Tower of Babel:" the two sides
were speaking different languages. The second
reason lies with the radical policy measures
proposed by certain writers, notably Thornstein
Veblen and the Technocrats. As pointed out above,
Veblen targeted most of his criticism at the "for-
profit economy," the cornerstone of mainstream
economics. While a virtue in classical and
neoclassical political economy, it (i.e. profit
motive) was a vice in the Economics of Abundance,
preventing the U.S. economy from attaining its new
found, higher potential. Clearly, the debate, like
most of 19th century political economy, had gone
beyond the realm of positive science, into the realm
of politics. In a true science, Veblen and the
Technocrats' hypotheses, however disturbing, would
have been subjected to empirical testing. For
example, did firms, as Veblen maintained, resist
lowering prices?

The third reason is its diagnosis of the Great
Depression. The Great Depression, Technocrats
argued, resulted from the U.S.'s inability to
translate potential abundance into actual abundance.
This, in turn, owed to either downward price
rigidity and/or deficient demand. Clearly, such
views were heretical, to say the least (Say's Law).
The fourth reason, and perhaps the most important,
was its policy prescriptions. Veblen and the
Technocrats, like Claude Henri de Saint-Simon in the

early 19th century, called for a radical reorganization of industry, the end of finance capitalism, and the emergence of technocracy. Finance capitalism was flawed, they argued. Profit maximizing managers would, accordingly, be replaced by efficiency- and output-maximizing technology-aware civil servants.

Merit notwithstanding, the Economics of Abundance posed a serious threat to mainstream political economy, itself the outgrowth of an earlier political debate, and, perhaps more importantly, to society in general. Like Robert Owen a century before, Veblen and the Technocrats questioned the ability of a private, for-profit economy to produce at capacity.

Not surprisingly, the Economics of Abundance, born of energy and controversy, failed to make inroads into mainstream political economy. Today, nothing of it remains, not the role of energy in production, not the estimates of U.S. potential output in the early part of the century, not its views of downward price rigidity in the face of technological change. The reason why is painfully obvious: to accept or to agree with any part of Technocracy was to accept all of it, lock, stock and barrel.

A good example of this is found in a "public policy pamphlet" by Aaron Director of the University of Chicago, entitled "The Economics of Technocracy" (Director 1933). Riddled with sarcasm, The Economics of Technocracy was an attempt by a leading member of the U.S. economics profession to "debunk" the proposed new science of wealth. Take, for example, the opening sentences of the Introduction, written by University of Chicago Press Public Policy Pamphlets series editor Harry D. Gideonese: "Labor-saving machinery saves labor. This startling discovery of the technocrats has led them to important conclusions" (Director 1933, 1). On page 4, the author, Aaron Director, attempts to discredit Howard Scott, one of Technocracy's founding members, by invoking his past ties to the I.W.W. (International Workers of the World). Turning to more substantive issues, he summarizes Technocracy in terms of a series of six points.

> 1. The importance of energy: "Through the expenditure of energy we convert all raw materials into products that we consume and through it operate all the equipment that we use." This, of course, has always been familiar to us, except that it was stated in terms of work, and not of energy. The great merit of the latter term is the possibility of dragging in the Law of Conservation of Energy and this marrying physics to the social mechanism.
> 2. Energy can be measured, and the unit of

> measurement is always the same, while the
> dollar varies from time to time.
> 3. The chief distinction between our society
> and that of all previous societies is the much
> greater amount of energy which can be
> generated. This has always been recognized by
> the designation of our civilization as the
> machine era.
> 4. With every increase in the amount of
> mechanical energy the need for labor decreases.
> 5. The present depression marks the end of an
> era, since the increase in mechanical energy
> has at last become so great that, regardless of
> what happens, the need for human labor will
> rapidly decline.
> 6. Does it follow, therefore, that the price
> system must break down, and that only the
> engineers can run a mechanical civilization.
> (Director 1933, 8)

He then proceeded to reexamine, using standard neoclassical analysis, each of these points. In keeping with the 19th century tradition of equating energy with machinery, he spoke in terms of "technical progress," and not of energy deepening. This is then followed by a Ricardian-inspired analysis of the effects of "technical progress" on costs, wages and prices. Competition, he argues, is a sufficient condition for full employment.

> On the other hand, the technocrats imply that a
> more scientific utilization of existing
> equipment would result in a much larger
> product. It is only necessary to insist that
> the number of engineers in industry far
> outweigh the number of economists, and if these
> engineers are to run industry in the future,
> they should be competent to point out methods
> of improving efficiency. It is not enough to
> hide behind a barrage of words. It should be
> patent to the most critical observer that the
> one thing which the individual enterprise under
> competitive conditions does strive for is to
> reduce its cost, regardless of the consequences
> on employment. (Director 1933, 16)

Having concluded that "technical progress is not incompatible with full employment," he proceeds, in Chapter VII, to debunk the view that the Great Depression was the result of energy-based technological change. This, metaphorically speaking, is where the gloves come off. First, he, in the tradition of Say and Ricardo, rules out underincome. Output, he argues, is identically equal to income, whether in the form of money or in kind.

> If there were no commercial banking system, the
> national income would be distributed for
> consumption goods and the production of
> additional equipment in accordance with the
> desires of the community. The output of
> industry is equal to the income of the laborers
> employed in it and of the property owners whose
> capital is invested in it.   Clearly, if
> entrepreneurs borrowed funds directly from the
> income receivers, they could not continue to
> produce capital equipment in excess of the
> amount which income receivers were willing to
> save. (Director 1933, 21)

In short, according to Director, Technocracy
offers nothing new, and, more importantly, is
riddled with the most elementary of oversights and
errors.   Energy is nothing new, and, more
importantly, presents no particular challenge to
mainstream political economy.   Technological
progress, in this case, electric drive, increases,
in a commensurate fashion, income, wages and
profits. The causes of the Great Depression, he
argues, lie elsewhere, notably in "the war, the
resulting debts, and tariffs."
Despite its ignominious end, the various themes
found in the Economics of Abundance continued to
generate interest in political economy for years
after, vindicating, as if it were, its scientific
merit.   For example, the role of energy in
production and growth lived on after World War II in
the form of growth accounting in general, and the
"Solow residual" in particular (Beaudreau 1995a,b,c,
1998). The role of downward price rigidity, as most
know, can be found in the work of Gardiner Means
(1935), John Maynard Keynes (1936), Paul Sweezy
(1939), and numerous others (including recent New-
Keynesian macroeconomics (Mankiw and Romer 1991).
The first half of the 20th century saw things
deteriorate further for political economy. In fact,
one could go as far as to argue that were it not for
Keynesian political economy with its user-friendly
policy measures, there is every reason to believe
that a major upheaval in social and economic
organization would have occurred. Classical and
neoclassical political economy were in shambles.
The Great Depression remained an enigma. Many
questions, few answers.
As we have attempted to show here, the gap
separating theory from reality, already substantial,
continued to widen. Belief in the virtues of the
market as an equilibrating mechanism was shattered.
Unemployment had become chronic, or so it seemed.
High-throughput, mass-producing firms had chosen not
to "flood" the market (?) and, in the process,
provoke a generalized fall in product prices.

Instead, as Veblen pointed out, they adjusted supply to demand, thus confirming the non-Walrasian nature of product markets.

By now, the problem was autoregressive in nature, dating back to the early classical political economists' failure to identify, model, and analyze the myriad aspects of inanimate energy-powered, high-throughput continuous-flow production processes. Neoclassical political economy, while successfully fending off radical political economists, failed to make significant advances in the study of such processes. Early 20th century political economists, in America and on the Continent, were then called upon to study what has to rank as one of the greatest quantitative innovations in the history of material civilization, namely the electrification of industry, with a set of analytical tools and constructs best suited for studying, at best, the production processes of Neanderthals and Cromagnons.

As we have tried to show in this chapter, we are not the first to make such an observation. In the 1920's and 1930's, at the height of the Great Depression, scientists and engineers called for a wholesale revamping of political economy, one based on energy. The Continental Committee on Technocracy, dismissed outright by political economists, stands as a testimony to the man's desire for truth, in spite of its many detractors. Clearly, something was terribly wrong.

While political economy continued to lack an empirically-consistent model of production, the years which followed the Great Depression were nonetheless rich in theoretical developments, the large majority of which relating to high-throughput production processes. Most of these were in the area of industrial organization. Moreover, most were aimed at re-examining the workings of markets in the presence of a limited number of high-throughput producers.

In the next section, we examine the post-WWII period, one characterized by massive energy deepening in the form of increased electric power consumption per lower-level supervisor (worker). We show how the dialectic between neoclassical political economy and various revisionists continued unfettered. For example, it will be argued that growth accounting was a response to this. Output and productivity, U.S. political economists noted, were far outdistancing growth in capital and labor, raising the myriad questions which dominate real business cycle analysis today. Productivity was increasing without a cause, or so it appeared.

The Second-Second Industrial Revolution

The second-second industrial revolution (post-WWII era) differed fundamentally from the first. Whereas the first operated on the extensive margin, increasing the breadth of material processes driven by electric power, the second operated on the intensive margin as machine speeds increased as a result of continued energy-deepening. Alfred Chandler best described this process in his perspicacious account of the managerial revolution in American business:

> In modern mass production, as in modern mass distribution and modern transportation and communications, economies resulted more from speed than from size. It was not the size of the manufacturing establishment in terms of number of workers and the amount and value of productive equipment, but the velocity of throughput and the resulting increase in volume that permitted economies that lowered costs and increases output per worker and per machine. (Chandler, 1977, 244)

In the decade from 1945 to 1955, total industrial consumption of electric power more-than-doubled going from 146,261 million kilowatt hours in 1945 to 334 to 334,088 million kilowatt hours. From 1955 to 1965, it increased by another 60 percent. By 1965, the United States consumed more electric power than all other countries combined. Never before in the history of mankind had per-capita energy consumption increased so rapidly.

This "energy orgy" came to a sudden, unexpected end with the oil/energy crisis in the 1970's. Reacting to the falling real oil prices, members of the Organization of Petroleum Exporting Countries (OPEC) orchestrated an artificial crisis in 1973, resulting in the quadrupling of oil and other energy prices. With this came the end of what had been over a century of declining real energy prices (Jorgenson 1983). Per-worker energy consumption, monotone increasing for most of this century, stopped rising, resulting in zero productivity growth.

This section examines the massive energy deepening which characterized the second-second industrial revolution. As was the case in the mid-to-late 19th century, there was no new energy form or transmission technology per se; instead, the emphasis was on electric power-based enabling technologies that resulted in massive energy deepening, defined as an increase in the consumption of a given type of energy—in this case, electric power—per unit time. As pointed out earlier, the 19th century witnessed a marked increase in horsepower per factory worker (see Table 3.5) the result of various innovations in steam engine-based enabling technologies (e.g. high-pressure steam engines, steam turbines).

Electric Drive-Based Energy Deepening and Manufacturing Productivity

By the end of World War II, electric-motors had supplanted belting, shafting and gearing as the dominant energy transmission technology (Devine 1990; Nye 1990). Fossil fuels and hydraulic resources, however, remained the ultimate power source, in the United States, Germany and Japan, as well as in most other industrialized nations. While in the early part of the century, most manufacturing firms generated their own electric power, by the end of WWII, most were purchasing it from publically-regulated electric power utilities. Referring to Table 4.1, we see that in 1945, electric utilities' installed generating capacity stood at 50,111,000 kilowatts, which, by 1955, had more than doubled to 114,472,000 kilowatts. By 1965, it had, once again, doubled, reaching 236,126,000 kilowatts. U.S. industrial generating capacity in this period increased, but at a considerably slower rate, going from 12,757,000 kilowatts in 1945 to 18,973,000 kilowatts.

Industrial power consumption increased at record rates in this period. Production processes were speeded up; throughput rates increased, resulting in higher plant output per period of time. What is particularly remarkable is the fact that most of this occurred with the existing capital and labor. For example, in the pulp and paper industry, existing paper machines turned faster, increasing productivity (Anderson, Bonsor and Beaudreau 1982).

In earlier work (Beaudreau 1995a,b,c,d,1996a,1998), we modeled what Chandler's "velocity of throughput" using a modified KLEMS production function, the KLEP (capital, labor and electric power) production function, and proceeded to estimate the relevant output elasticities directly using post-WWII input and output data for U.S., German and Japan manufacturing. These were then used in lieu of factor shares as the relevant weights in standard growth accounting exercises. Specifically, we assumed the existence of a well-behaved, twice differentiable, monotonic and quasi-concave production function, such as (1), where Q = value added, EP = electric power, L = employment, and K = capital. Christensen and

Jorgenson (1970) and Gollop and Jorgenson (1980) define the rate of growth of total factor productivity, tfp = dTFP/dt 1/TFP, as tfp = q-s_{EP} ep - s_Kk - s_Ll, where q = dQ/dt 1/Q, ep = dEP/dt 1/EP, l = dL/dt1/L, and k = dK/dt 1/K, and s_i is the weighted average of the ith factor share over the discrete time interval $\forall$ we = EP, L, K (Gullickson and Harper 1988).

$$VA(t)=F[EP(t),L(t),K(t)] \qquad (3.1)$$

The relevant output elasticities were estimated directly (i.e. as opposed to indirectly using the cost function).[6] Specifically, data on value added, electric power consumption, total employment and capital for U.S., German and Japanese manufacturing were used to estimate the Cobb-Douglas KLEP production function: $Q = EP^{\beta_1} L^{\beta_2} K^{\beta_3}$.

Table 3.7
KLEP Output Elasticities: U.S., German and Japanese Manufacturing

Independent Variables	U.S. 1950-1984	Germany 1963-1988	Japan 1965-1988
EP	0.537244	0.747482	0.605599
	(26.551)	(3.135)	(3.017)
L	0.399727	0.121134	0.197653
	(18.231)	(2.332)	(1.847)
K	0.075049	0.131383	0.196748
	(2.768)	(0.543)	(1.608)
Constant	0.075049	0.046106	-0.019274
	(9.956)	(1.426)	(0.271)
R^2	0.98438	0.95821	0.98314
F	1008.5140	229.2853	612.1780

The estimated output elasticities for all three countries are presented in Table 3.7. What is particularly striking are the similarities across countries. In all three cases, electric power consumption is, by far, the most important factor input, as evidenced by output elasticities of 0.537244, 0.747482 and 0.605599.

Pre- and Post-Energy Crisis Growth

Table 3.8 reports the relevant growth rates for manufacturing value added (Q), electric power

(EP), labor (L) and capital (K), as well as the
relevant fixed-weight aggregate input index (AI)
(Hisnanick and Kymm 1992; Gullickson and Harper
1988) for three time intervals: 1950–1984,
1950–1973, and 1974–1984.

Table 3.8
Output and Input Growth Rates: U.S., German and Japanese
Manufacturing

U.S.

	1950-1984	1950-1973	1974-1984
USVA	2.684	3.469	0.121
USAI*	2.674	3.472	0.310
USEP	4.052	5.371	0.246
USN	0.662	0.900	-0.091
USK	3.694	3.614	3.4008

Germany

	1963-1988	1962-1973	1974-1984
GERVA	2.462	6.522	1.486
GERAI*	2.433	5.190	1.080
GEREP	2.894	5.883	1.366
GERN	-0.785	0.592	-0.938
GERK	2.945	5.620	1.406

Japan

	1965-1988	1965-1973	1974-1988
JAPVA	3.826	8.844	3.099
JAPAI*	3.566	9.856	1.538
JAPEP	3.559	11.320	0.965
JAPN	-0.082	2.297	-0.367
JAPK	7.520	13.536	5.182

*$\hat{\beta}_1 \frac{ep}{ep} + \hat{\beta}_2 \frac{i}{i} + \hat{\beta}_3 \frac{k}{k}$ where $\hat{\beta}_i$'s are the estimated coefficients.
Source: Beaudreau (1998), 99.

Unlike previous studies that found a non-
negligible gap between rates of output growth and
rates of aggregate input growth (i.e. the Solow
residual), these results indicate that growth in
U.S., German and Japanese manufacturing value added
is fully explained by growth in the relevant fixed-
weight factor input growth indexes (Beaudreau
1995a,b,c,1998). For the complete period 1950
1984, U.S. manufacturing value added increased at an
average annual rate of 2.684 percent, while the
aggregate input increased at 2.655 percent. For the
first sub-period 1950 1973, it increased at 3.469,
while the aggregate input increased at 3.466
percent. In the second sub-period, 1974 1984, it
increased at an average annual rate of 0.121, while
the aggregate input increased at 0.310.

Chief among the causes of growth in U.S., German and Japanese manufacturing value added is electric power consumption. In the U.S. case, electric power consumption increased at an average annual rate of 4.052 percent over the period 1950 1984. Per worker consumption of electric power in U.S. manufacturing in this period goes from 12,534 kilowatt hours in 1950 to 41,688 kilowatt hours in 1984, a total increase of 232 percent (U.S. Department of Commerce, various years). When multiplied by the relevant output elasticity (i.e. 0.537244), the growth in electric power consumption in U.S. manufacturing accounts for 82 percent of overall output growth (i.e. 2.17/2.68), which corroborates the predictions of E-0 production analysis. Prior to the energy crisis (i.e. 1973), electric power consumption in manufacturing increased at an average annual rate of 5.371 percent. Output increased at an average annual rate of 3.466 percent. Electric power consumption growth, it then follows, accounts for 83 percent of output growth (i.e. 2.88/3.46). In the post-energy-crisis period, electric power consumption increased at an average annual rate of 0.246, while output increased at an average annual rate of 0.121.

Pre- and Post-Energy Crisis Productivity Growth

Gullickson and Harper (1988) and Hisnanick and Kymm (1992) examined the sources of productivity growth in U.S. manufacturing using $lp = q - l$ (see (Equation 3.2)) as the appropriate measure of labor productivity.[2] In this section, we report a series of revised estimates of the role of electric power and capital in labor productivity in U.S., German and Japanese manufacturing. Referring to Table 3.9, we see that the rate of growth of labor productivity is, in all three cases, entirely explained by increasing electric power and capital intensities, defined here as $ep - l$, the shift away from labor to electric power, and $k - l$, the shift away from labor to capital.

$$lp = q - l = \beta_1 [ep - l] + \beta_3 [k - l] + tfp \quad (3.2)$$

Over the entire sample period (i.e. 1950 1984), U.S. labor productivity increased at an average annual rate of 2.022 percent. During this period, the electric power-labor ratio increased at an average annual rate of 3.39 percent, which when multiplied by the relevant elasticity (i.e. 0.537244), yields a value of 1.820 percent, which measures the effects on labor productivity of the substitution of labor for

electric power referred to by Dale Jorgenson
(Jorgenson 1988). Also, during this period, the
capital-labor ratio increased at an average annual
rate of 3.032, which when multiplied by the
relevant output elasticity (0.075049) yields a value
of 0.191, which measures the effects on labor
productivity of the substitution of labor for
capital. The sum of these two effects accounts for
99.5 percent of the growth in labor productivity in
U.S. manufacturing (2.011/2.022).

Prior to the energy crisis, labor productivity
increased at an average annual rate of 2.569
percent, of which 2.400 percent can be
attributed to labor-electric power substitution
(energy deepening) and 0.170 percent can be
attributed to labor-capital substitution.
Together, these two effects overstate the overall
increase in labor productivity by 1 percent. In the
ensuing decade, labor productivity increased at an
average annual rate of 0.212 percent, of which
0.180 percent can be attributed to labor-electric
power substitution and 0.219 percent can be
attributed to labor-capital substitution.

Table 3.9
Productivity Growth: U.S., German and Japanese
Manufacturing

U.S.

Period	1950-1984	1950-1973	1974-1984
uslp	2.022	2.569	0.212
ustfp	0.010	-0.002	-0.1889
β_1(usep-usn)	1.820	2.400	0.180
β_3(usk-usn)	0.191	0.170	0.219

Germany

Period	1963-1988	1963-1973	1974-1988
gerlp	3.247	5.930	2.424
gertfp	0.007	1.315	0.3942
β_1(gerep-gern)	2.750	3.954	1.722
β_3(gerk-gern)	0.489	0.660	0.307

Japan

Period	1965-1988	1965-1973	1974-1988
japlp	3.908	6.547	3.466
japtfp	0.208	-1.126	1.567
β_1(japep-japn)	2.204	5.463	0.806
β_3(japk-japn)	1.495	2.210	1.091

Source: Beaudreau (1998), 101.

The Role of Capital and Labor in Growth Accounting

Traditional growth accounting, best associated with the work of Edward Denison, and E-O production analysis-based growth accounting (Beaudreau 1998) are opposites. The former, Paleolithic in its approach to modelling material processes, attributes growth to capital, labor, and a residual, while the latter attributes growth to energy consumption, and, innovations in second-law efficiency. The former views energy as an intermediate input and consequently not productive in the physical sense; the latter views capital and labor as organizational inputs, and, consequently, not productive in the physical sense.

Unless it can be shown that capital and labor are physically productive (i.e. are a source of energy) in modern production processes, reason dictates that the latter view should prevail, namely that capital and labor are organizational inputs.

E-OPA Growth Accounting

This simple finding has important implications for growth accounting in general. For example, as argued in Beaudreau (1998), it implies that the focus of growth accounting in general should be on two variables, namely energy and second-law efficiency, the latter being related to tools (capital) and upper- and lower-level supervision (labor). Tools and supervisors affect growth via η, second-law efficiency. Operationally, this implies that the basic E-O production function $(W(t)=\eta(t)E(t))$ should be the theoretical basis of growth accounting. The rate of growth of work $\dot{W}(t)/W(t)$, it therefore follows, should, by definition, be equal to the sum of the rate of growth of second-law efficiency $\dot{\eta}(t)/\eta(t)$, and the rate of growth of energy consumption $\dot{E}(t)/E(t)$. This is formalized in terms of Equation 3.3 below.

$$\dot{W}(t)/W(t) = \dot{\eta}(t)/\eta(t) + \dot{E}(t)/E(t) \qquad (3.3)$$

Thus, the rate of growth of work (i.e. value added) is an increasing function of the rate of growth of second-law efficiency and the rate of growth of energy consumption. Table 3.10 presents the corresponding values for U.S., German and Japanese manufacturing. Second-law efficiency was defined as the ratio of value added to electric power consumption. We see, for example, that in U.S. manufacturing value added increased at a rate of 2.5625 percent from 1950 to 1984. Electric power consumption in this period increased at a rate of 3.9342 percent, while second-law efficiency decreased at a rate of 1.3203 percent. Manufacturing firms in all three countries were less able to extract work from each additional kilowatt hour (i.e. lower energy). Broken down into pre- and post-energy crisis sub-periods, we see that in both, second-law efficiency decreased, albeit at a slower rate in the latter period. The corresponding values for German and Japanese manufacturing show that while second-law efficiency decreased from 1964 to 1988, it increased in Germany. Moreover, while the energy crisis failed to arrest the decline in second-law efficiency in U.S. manufacturing, it succeeded in Japanese manufacturing, where it went from an average annual decrease of 2.1835 from 1964 to 1973, to an average annual increase of 1.12975 from 1974 to 1988.

Table 3.10
E-O Analysis-Based Output and Input Growth Rates: U.S., German and Japanese Manufacturing

U.S.

	1950-1984	1950-1973	1974-1984
USVA	2.684	3.469	0.121
USEP	4.052	5.371	0.246
$US\eta$	-1.367	-1.902	-0.124

Germany

	1963-1988	1962-1973	1974-1988
GERVA	2.462	6.522	1.486
GEREP	2.894	5.883	1.366
$GER\eta$	-0.432	0.638	0.066

Japan

	1965-1988	1965-1973	1974-1988
JAPVA	3.826	8.844	3.099
JAPEP	3.559	11.320	0.965
$JAP\eta$	0.267	-2.479	2.134

Source: Beaudreau (1998), 103.

These results are self-explanatory. The phenomenally-high, uninterrupted growth in U.S., German, and Japanese manufacturing in the post-WWII period was the result of energy deepening in the form of increasing levels of electric power consumption per period of time. Factories turned out more and more output, the result of more and more electric power, confirming Alfred Chandler's views on productivity growth and its causes. This,

however, came to an abrupt end in the 1970's with
the "energy crisis," as firms curtailed energy
deepening. Energy conservation became the norm.
Electric power consumption growth fell dramatically,
as did growth and productivity. The energy orgy was
over, or so it appeared.

Electric Drive-Based Energy Deepening As Seen by Political Economists

By the end of World War II, mainstream political
economy was in crisis. None of the received wisdom
(i.e. theory) fit the data, not price theory, not
wage theory, not monetary theory, etcetera.
Contributing to this state of crisis was the
mercurial rise of Keynesian political economy (macro
and micro). Traditional political economists,
Marshallian in their approach to the world, had to
deal with the General Theory and its many
ramifications. Would government (e.g. the Full-
Employment Act of 1946) supplant the market? Had
markets failed? Were Marx, Engels and other 19th
century radicals right after all? Was capitalism
inherently flawed? Ideology had returned with a
vengeance.
Breaking the fall, metaphorically speaking, were
the high growth rates that characterized this
period. As shown above, growth rates reached new
heights in the post-WWII period (1945 1973).
Clearly, while something had gone terribly wrong in
the 1920's and 1930's, the unprecedented growth of
the 1940's, 1950's and 1960's had more than made up
for it. Per-capita output increased monotonically
over the course of this period, until, of course,
the energy crisis.
In this section, we examine electric drive-based
energy deepening—and its end in the 1970's—as seen
by political economists. It will be argued that
unlike the early 20th century, which had witnessed
the development of new analytical approaches to
production based on energy (Technocracy, The
Brookings Institution approach to production, Social
Credit, Frederick Soddy's Cartesian economics), the
post-WWII period, despite being a period of
unprecedented energy deepening, was devoid of any
such developments.
This only serves to beg the question why? Why
was it that in the midst of an energy revolution,
political economy, the purported "science" of
wealth, failed to even identify energy in general,
and electric power in particular, as a key cause of
growth? The answer, we believe, is "bad" science.
As pointed out above, the post-WWII period was
one of crisis for the young science of political

economy. On the one hand, output was growing at unheard of—and unexplained—rates. Clearly, production theory had failed to shed light on the causes, the result of which was the birth of Moses Abramovitz's "measure of ignorance" (Solow residual). On the other hand, the Great Depression, and the chosen remedy, Keynesian-style government expenditure policies, continued to stir emotions in what had been throughout the 19th century, a political ideology, namely market-based political economy. The new debate pitted classical and neoclassical political economists (under the auspices of monetarism) against Keynesian political economists. At stake were a number of issues, including the causes of the Great Depression, the wisdom of the chosen solution (New Deal), the wisdom of the Full Employment Act of 1946, and the role of government in the economy. For three decades, political economists, whether of the classical/neoclassical or Keynesian persuasion, engaged in rhetorical debates, basing their arguments on a Paleolithic model of production—outdated by over three centuries—and a Paleolithic model of exchange, an example of which is Don Patinkin's Money, Interest and Prices, published in 1965.

Bad science, we contend, is also to blame for the rise of formalism in post-WWII political economy. A good example is production theory, where the Paleolithic-era model of production referred to above was metaphorically cloaked in a series of new suits, including the Constant Elasticity of Substitution suit, the Translog suit, the Leontief suit, and the Cobb-Douglas suit. Throughout this period, formalism became (and continues to be) synonymous with "good" science. Today, professional journals are replete with formalizations of antiquated, Paleolithic production and exchange technologies. It comes as little surprise that paradoxes and ironies have become the norm in political economy, not the exception.

Electric Drive-Based Energy Deepening as Seen by Political Economists

While they ignored energy and the role of energy in the creation of wealth, political economists could not ignore the phenomenal growth in output in general, and output per worker (conventionally-defined labor productivity) in this period. Among the first to acknowledge the existence of a divergence between conventionally-used factor inputs (capital and labor) and output were Jan Tinbergen in Europe and Moses Abramovitz in the United States.

Both had devised indexes of input and output growth, and found, using post-WWI data, the presence of an unexplained, upward drift in output indexes. Abramovitz referred to this discrepancy as "our measure of ignorance."

> Credit for the earliest explicit calculation belongs to Tinbergen, who in a 1942 paper, published in German, generalizes the Cobb-Douglas production function by adding an exponential trend to it, intended to represent various "technical developments". He computed the average value of this trend component, calling it a measure of "efficiency", for four countries: Germany, Great Britain, France, and the U.S., using the formula t = y -2/3n - 1/3k, where y, n, and k are the average growth rates of output, labor and capital respectively, and the weights are taken explicitly from Douglas.
>
> Note how close this is to Solow who will let these weights change, shifting the index numbers from a fixed weight geometric to an approximate Divisia form. Nobody seems to have been aware of Tinbergen's paper in the U.S. until much later.
>
> The developments in the U.S. originated primarily at the NBER where a program of constructing national income and "real" output series under the leadership of Simon Kuznets was expanded to include capital series for major sectors of the economy, with contributions by Creamer, Fabricant, Goldsmith, Tostlebe and others. (Griliches 1995, 3)

Unfortunately, in spite of half of century of work on growth since, our "measure of ignorance" has not decreased (Maddison 1987; Gullickson and Harper 1988; Aghion and Howitt 1998). It has changed names, been cloaked in formal and informal models, been tested against every possible datum, yet, in the end, it stands as a living testimony to what by now were over two centuries of "bad" science.

Examples continue to abound in the professional journals. Take, for example, a recent paper in The Quarterly Journal of Economics by Walter Nonneman and Patrick Vanhoudt, entitled "A Further Augmentation of the Solow Model and the Empirics of Economic Growth for OECD Countries," where production is modeled in terms of the following functional form.

$$Y_t = cL_t^{(1-\Sigma\alpha)} K_{1t}^{\alpha_1} K_{2t}^{\alpha_2}, \ldots , K_{mt}^{\alpha_m} \qquad (3.4)$$

L is labor, and K_i is capital of type i. There are, altogether, m types of capital, including infrastructure, equipment, other physical capital,

human capital, and know-how. Clearly, political
economists are, as far as modeling production is
concerned, no further today than they were 200 years
ago. All of Nonneman and Vanhoudt's inputs are
organizational inputs, affecting productivity only
indirectly through η, the level of second-law
efficiency. Accordingly, improvements in any or all
of these will affect output, but only marginally.

The Energy Crisis and Productivity as Seen by Political Economists

As pointed out earlier, the energy "orgy" of the
post-WWII period came to an unexpected and
unfortunate end in the mid-1970's with the OPEC-
induced energy crisis. Energy prices doubled,
tripled, and, eventually, quadrupled, pushing the
West up its demand curve for energy. Energy
consumption stopped growing. In hindsight, one
would think that, at long last, energy would get the
attention it deserved. "Absence makes the heart
grow fonder," as the saying goes. Figuratively
speaking, the hen that had laid the golden eggs
(energy rents) for over two centuries was no more.
As incredulous as it may appear, this was not to be,
as evidenced by the continued absence of energy from
production theory, my work (Beaudreau
1995a,b,c,d,1998) notwithstanding. Once again, the
reasons are many. First, since energy had, up to
then, not been defined as a factor input, the focus
as far as production theory was concerned was not on
energy per se, but rather, on the relationship
between energy prices and the demand for capital and
labor. For example, if energy and capital were
found to complementary, then an increase in the
price of the former will decrease the demand for the
latter, reducing growth, and, quite possibly, in the
short run, provoking a recession (Berndt and Wood
1975).
To address these questions (complementarity-
substitutability), Ernst Berndt and David O. Wood
developed a new, all-inclusive production function,
commonly known as the KLEMS production function
(i.e. capital, labor, energy, materials, and
services). By eliminating the age-old distinction
between factor inputs and intermediate inputs,
Berndt and Wood cleared the way for the estimation
of energy-labor and energy-capital, not to mention,
energy-materials and energy-services, cross-
elasticities.
At first blush, one's inclination is to laud what
has to qualify as an important breakthrough, as far
as the role of energy in production is concerned.
After all, it took 200 years (from Smith's Wealth of

Nations in 1776 to 1975) for energy to break the
stranglehold held by capital and labor in production
theory. Finally, energy had entered the select club
of factor inputs. The irony, however, is that the
select club was (is) no more. The distinction
between intermediate input and factor input was
blurred beyond recognition. Nothing was a factor
input, and nothing was an intermediate input.

Further marring Berndt and Wood's contribution
was the method used to estimate the various
parameters (e.g. output and cross-elasticities).
Specifically, instead of using input and output
data, they chose cost and factor share data.
Implicit in the use of such data are the assumptions
of constant returns to scale in production and
competitive factor markets. It being the case that
electric power is the dominant form of energy in
manufacturing, and that the price of electric power
is regulated, doing so introduces an important bias.
Their results showed an energy output elasticity of
0.06 percent. Put plainly, energy was, for all
intents and purposes, insignificant. Case closed.

When it was suggested that the energy crisis
could lie at the root of the productivity and growth
slowdown, Berndt and Wood's energy output elasticity
(and others') were cited as evidence to the
contrary. Something as insignificant as energy in
20th century production processes could not possibly
have halved the growth rate, and all but eliminated
productivity growth, or so the story goes. In their
comprehensive study of total factor productivity in
U.S. manufacturing (the results are reproduced in
Table 3.11), William Gullickson and Michael J.
Harper, show, beyond a shadow of a doubt, that
energy was not the cause.

Table 3.11
Gullickson and Harper's Estimates of Multifactor
Productivity 1949 1983

Period	Output	Aggregate	Capital	Labor	Energy	Material	Service	KLEMS
1949 - 1983	3.1	2.0	3.8	0.8	3.3	2.2	4.6	1.1
1949 - 1973	4.2	2.7	3.9	1.5	5.1	3.1	5.4	1.5
1973 - 1983	0.6	0.3	3.6	-1.0	-0.8	0.2	2.6	0.3

Source: Gullickson and Harper (1987), 22.

The Energy Agnostics

Unconvinced by either the methodology or the results of growth accounting, others continued to study the role of energy in production. Among these, one finds political economists Dale Jorgenson, Zvi Griliches, and Nathan Rosenberg, and engineers-economists Sam H. Schurr, Calvin Burwell, Warren D. Devine Jr, and Sidney Sonenblum. For example, in 1983, Dale Jorgenson published the results of a study of the role of electricity in technical change (i.e. whether they be electricity using or not). Of the 35 sectors studied, 23 were found to be of the electricity-using type. Specifically, "the decline in electricity prices prior to 1973 prompted increased electrification via the substitution of electricity for other forms of energy, and through the substitution of energy for other inputs—especially labor (Jorgenson 1983, 21). Electrification and energy deepening in the form of higher levels of electricity consumption, he went on to conclude, played a fundamental role in productivity growth.

In similar work, Sidney Sonenblum shows that technological progres in manufacturing has been "related to the adoption and spread of production processes and modes of organization that are dependent on the use of electricity" (Sonenblum 1990, 277). The downside, however, in Sonenblum and Jorgenson's work is the absence of a theory of production sufficiently general to incorporate energy. Oddly enough, energy, in this case, electricity, affects output indirectly through "technology." Habits die hard. The spirit of the neoclassical production function with its emphasis on capital and labor, and, to a lesser degree, on technology, is very much present in Jorgenson's work, and, indeed, in all post-energy crisis work on energy.

This, in brief, is where the economics profession stands today on the role of energy in production. Despite two epoch-defining energy innovations, two centuries of energy deepening, and a cataclysmic "energy crisis," energy, the building block of the universe, remains very much the outsider, having no place at the table.

Conclusions

Perhaps the thing that stands out the most about the 20th century is the extent to which is resembles the 19th century in so far as technological change is concerned as well as the fallout. In both, new technologies led to massive energy deepening, unparalleled potential output growth and macroeconomic crises of unheard of proportions. Similarly, in both, the enabling technology spawned a series of sub-enabling technologies, related to the first but with specific consequences. Accordingly, the second industrial revolution can be deconstructed into two sub industrial revolutions, the first-second second-second industrial revolution. In the next chapter, we examine the fallout of the two energy crises of the 1970's. namely the end of second-second industrial revolution. The energy crises brought post-WWII energy deepening, characterized by rising energy to capital and energy to labor rations, to an end, bringing with it a decrease in productivity and output growth.

4

Growth without Growth: The Post-Energy Crisis Period

> In 1974, Robert Solow presented a model suggesting that a constant level of economic growth could be sustained indefinitely, in principle, provided that the elasticity of substitution of capital and labor (taken together) for exhaustible resources is greater than unity.... In fact, most of the theoretical work being done by economists at the time ignored the implications of the basic laws of thermodynamics. Economic theorists, at least briefly, seem to have reinvented the perpetual motion machine, an idea that seems very hard to kill.
>
> —Robert U. Ayres and Indira Nair,
> Thermodynamics and Economics

Introduction

In the previous two chapters, the analysis focused on "positive" energy-related innovations, be they a new power source, a new enabling technology, or energy deepening per se, as sources of industrial revolutions. In this chapter, we continue our look at industrial revolutions by examining what could be referred to as an industrial unrevolution, namely the end of the second-second industrial revolution. Throughout the post-WWII period, speed-up and other related enabling technologies had paved the way for massive energy deepening. The energy crises of the 1970's by increasing the cost of energy and the expected cost of energy brought this process to an end, with predictable results: productivity growth literally dried up.

The decision to include this "unrevolution" was based on purely scientific grounds. If our model of industrial revolutions is correct, then it should be able to explain not only the beginning but the end of an industrial revolution.

The chapter is organized as follows. First, using E-O production analysis and the bargaining model of distribution, we examine the "fallout" of the energy crisis, consisting of lower energy consumption growth, lower labor productivity growth, and the resulting "profit squeeze." As argued in earlier work (Beaudreau 1998), higher energy prices, in combination with higher real wages, "squeezed" profits in the 1970's and early 1980's.6 Profit-maximizing firms responded by (1) accelerating the rate of automation in the work place (replacing animate lower-level supervisors with inanimate ones), and (2) where not technically feasible, moving production offshore. Not surprisingly, this has contributed to higher earnings, and, consequently, higher share prices.

As in the previous chapters, this will be followed by an examination of the profession's response to these issues. Not surprisingly, it fares no better than previously. In fact, one could argue that it fares worse, given the absence of a "radical" fringe such as Technocracy.[1] As we shall show here, the recent emphasis on R D and education as growth panacea is mis-founded, as is the hype surrounding information and computers.

Post-1973 Output Growth

The centuries-old "energy orgy," characterized by ever-increasing levels of energy consumption, came to an unexpected end with the energy crisis of the 1970's and 1980's. While per-capita energy consumption remained high, it stopped growing, freezing per capita income, and consequently, the standard of living. As such, the current generation of lower- and upper-level supervisors (conventional labor) cannot expect to do better than the preceding one. Per capita income growth, predicated on per-capita energy consumption growth, has, for all intents and purposes, ended. Barring some unforeseen development in energy technology (e.g. the development of commercially-viable, environmentally-friendly energy from nuclear fusion), there is every reason to believe that low—in the limit, zero—energy consumption growth and consequently, low productivity growth, will be with us for a long, long time.

Output Growth in the Post-Energy Crisis Period

As Dale Jorgenson has argued, the energy deepening of the 20th century in the form of higher and higher electric power consumption per worker, was largely the result of a falling real price of electricity.

> Our empirical results strongly confirm the
> hypothesis advanced by Schurr and Rosenberg
> that electrification and productivity growth
> are related in a wide range of industries.
> Schurr et al. (1979) show that the price of
> electricity fell in real terms through 1971.
> This decline in real electricity prices has
> promoted electrification through the
> substitution of electricity for other forms of
> energy and through the substitution of energy
> for other inputs—especially for labor input.
> In addition, the decline in the real price of
> electricity has stimulated the growth of
> productivity in a wide range of industries.
> The spread of electrification and rapid growth
> of productivity through the early 1970s are
> both associated with a decline in real
> electricity prices. This decline was made
> possible in part by advances in the thermal
> efficiency of electricity generation.
> (Jorgenson 1984, 21)

Throughout this period, falling real prices prompted process engineers in these industries to devise new, more energy-intensive production processes, the defining feature of which was what Alfred Chandler refers to as "higher velocity of throughput" (speed). Production processes were speeded up, with the result that conventionally-defined productivity increased monotonically in the period. As such, lower- and upper-level supervisors were called upon to supervise increasing quantities of energy, and, consequently, of work. This "idyllic" situation came to a sudden and unexpected halt with the energy crises of the 1970's. Energy prices stopped falling and began rising, with the expected results on developments in production technology. Table 4.1 presents pre- and post-energy crisis electric power consumption growth rates for the U.S., Germany and Japan. In short, energy deepening, the principal cause of production-line speed-ups, had ended. More important, however, is the fact that, from this point on, energy conservation (energy shallowing) became the overriding concern in process engineering.

Table 4.1
Electric Power Consumption Growth Rates: U.S., German and
Japanese Manufacturing

U.S.				
	USEP	1950-1984 4.052	1950-1973 5.371	1974-1984 0.246
Germany				
	GEREP	1963-1988 2.894	1962-1973 5.883	1974-1988 1.366
Japan				
	JAPEP	1965-1988 3.559	1965-1973 11.320	1974-1988 0.965

Source: Beaudreau (1998).

The U.S. Pulp and Paper industry is a case in point. Higher energy prices, and, more importantly, the uncertainty over future energy prices have altered fundamentally the role of energy deepening in this industry, as evidenced by its post-energy crisis obsession with energy efficiency/energy. In his introductory remarks to a TAPPI (The American Pulp and Paper Institute) volume devoted to "Energy Engineering and Management in the Pulp and Paper Industry," editor Matthew J. Coleman explains:

> The basic resources needed to make paper are wood, water and energy. While fiber sources can change, it is not practical to make paper without water and energy. Electricity, coal, and fuel oil are the primary purchased sources of energy, with bark and hog fuel the primary indigenous sources of fuel available to mills. Whether a mill purchases all its power, or is able to generate its own using waste for fuel, the large amounts of energy used in the pulping and papermaking process have maintained the need for efficient energy management.
>   In North America during the past decade, the pulp and paper industry has made great strides toward energy independence. The availability of waste and bark for boiler fuel has focused the strategies of fully-integrated mills into more efficient use of these energy resources, while for non-integrated mills and many European mills, the management emphasis has been on better energy efficiency and conservation.
>   Mills which practice the best methods of energy recovery and management will benefit the most from the current instabilities in the world oil markets. The current price rise of oil is a reminder of how delicate the balance is between production and consumption of that particular fuel. Price increases in fuel oil

> can trigger increases in electrical power cost,
> and increase the substitution value of fuels
> such as coal, bark, and waste wood. (Coleman
> 1991, I)

Evidence that the energy crisis and its
aftershocks, especially the lingering uncertainty,
have affected production processes in the pulp and
paper industry is provided by Heikko Mannisto, vice
president of EKONO Inc., Bellevue Washington, who,
in the same volume, describes just how energy price
increases have "drastically" changed mill design.

> Because of the energy crisis in the early
> 1970's, energy conservation became one of the
> key issues in the North American pulp and paper
> industry, and most companies initiated millwide
> studies on how to reduce energy consumption.
> Because the mills had been designed and built
> for low energy costs, there were suddenly more
> conservation measures that had a high or
> reasonable payback than many companies could
> manage. (Mannisto 1991, 21)

As this example clearly shows, the emphasis in
manufacturing (U.S., German, and especially
Japanese) is now on maximizing second-law efficiency
(η), not maximizing throughput. Theoretically
speaking, this involves lowering energy consumption
levels in production processes closer to their
theoretical minimums. More energy-intensive
production processes (i.e. higher throughput) are,
in spite of lower real energy prices, no longer an
option, owing in large measure to the pervading
uncertainty over energy prices. That energy prices
have fallen in real terms over the past decade is
immaterial. The emphasis is, and no doubt will
continue to be, on increasing energy efficiency.

The Productivity Slowdown as Seen by Political Economists

How has the economics profession responded to
the "productivity slowdown" as described here.
Before addressing this question, a few remarks are
in order. First, it bears reminding that the problem
of growth is as old as political economy itself.
Adam Smith's primary concern, as evidenced by the
title of his magnum opus, was with the "causes of
the wealth of nations." Few, however, were those
who followed in his footsteps. Classical, radical,
and neoclassical political economists were, as shown
earlier, more concerned with distribution than with
wealth production. Likewise, monetarists and
Keynesians in the post-WWII period were more

concerned with full employment than with wealth production per se. This indifference, however, came to an end with the recent "wildfire interest" in economic growth.

This, we maintain, is essential to understanding the profession's response. First, the paucity of work on growth in the past has made for the current situation in which surprisingly little is known. This can, in part, be attributed to the "parametric" view of technology that has characterized production theory from William Cobb and Paul Douglas' first analytical production function in 1927. Technology is determined exogenously, or so the profession has maintained.

As growth rates in the mid-to-late 1980's failed to return to their pre-1973 levels, it became painfully obvious that our knowledge of the A in AK approach was inadequate, and in need of a major overhaul. The profession responded in two ways. The first was in terms of what are today referred to as AK models, popularized by Paul Romer (1986) and Robert Lucas (1988). The second was in terms of Schumpeterian models of innovation (Aghion and Howitt 1998).

Implicit in both is the view that technology—at least in so far as aggregate growth is concerned—is a social phenomenon. Growth rates in Western industrialized economies are not the result of an individual process or product innovation, but of an agglomeration of such innovations, each having a common cause, whether it be higher education, more basic research, or novel "General Purpose Technologies." Both approaches should as such be seen for what they are, namely attempts at mapping the social DNA of innovation.

In their work on innovation hierarchies, Elhanan Helpman and Manuel Trajtenberg describe "General Purpose Technologies," as follows:

> In any given "era" there typically exist a handful of technologies that play a far-reaching role in fostering technical change in a wide range of user sectors, thereby bringing about sustained and pervasive productivity gains. The steam engine during the first industrial revolution, electricity in the early part of this century and microelectronics in the past two decades are widely thought to have played such a role. (Helpman and Trajtenberg 1994, 1)

Still in its early infancy, this literature has yet to generate testable, and indeed, tested predictions. In fact, it is my view that this literature, being based on a Paleolithic model of

production, will fail to produce any important
breakthroughs. As we have maintained throughout
this book, political economy has, for over two
hundred years, worked with an archaic model of
production. Capital and labor are not productive
factor inputs, but, rather, are organization inputs,
affecting η, the level of second-law efficiency.
Until these simple facts, common knowledge in
production engineering and in elementary physics,
are integrated into production theory, there is
little hope that the work now under way will explain
the past, understand the present and predict the
future. In other words, there is little hope that
these models will shed light on "the" issue of the
1980's and 1990's, the productivity slowdown.

Automation and Freer-Trade as Seen by Political Economists

We begin with automation. At the risk of
sounding repetitive, the main problem in so far as
the literature on information-based technologies is
concerned, is the absence of a model of production
sufficiently general to include, in a meaningful
way, information as a factor of production. As it
now stands, information is lumped into what is
commonly referred to as "technology," the A in AK.
More information, it therefore follows, renders
capital—including human capital—and labor more
productive.

This is best illustrated by recent work on
computers and productivity. A good example of this
is Lehr and Lichtenberg's recent work on computers
and productivity growth in U.S. federal government
agencies. Not knowing just how to handle computers
analytically, they do what "neoclassical apologists"
have done for decades, namely, include them as a
separate input in production functions, and proceed
to estimate the relevant output elasticities. They
find a value of 0.06 for the computer-output
elasticity and interpret it as evidence that
computers are, in fact, productive.

Further, the absence of an empirically-consistent
model of production has distorted the debate over
the role of information and automation at the firm
and at the aggregate level. As pointed out above,
since the end of energy deepening (1973 1980), firms
have resorted to automation and off-shore production
to increase profits and earnings. Most engineers
know perfectly well that information has never been,
is not, and never will be productive in the physical
sense. They also know that labor costs constitute
(constituted) the single most important cost factor,
and that assuming a 65/35 wage/profit split, a one

percent decrease in labor costs results in a 2 percent increase in profits.

In spite of this, there is a widely-held view that automation and information will, eventually, lead to higher productivity. Professor Paul David's view on information and productivity is a case in point. Responding to the apparent failure of "new information technologies" to increase growth and productivity, he pointed to the presence of significant diffusion lags. In "The Dynamo and the Computer: An Historical Perspective on the Modern Productivity Paradox," he shows that the effects of another important innovation, the electrification of industry in the early part of the century, were not instantaneous, but rather were diffused over a long, somewhat protracted period. The effects of the information revolution now upon Western industrialized economies, he surmises, will be felt over time.

> But, even these cautionary qualifications serve only to further reinforce one of the main thrusts of the dynamo analogy. They suggest the existence of special difficulties in the commercialization of novel (information) technologies that need to be overcome before the mass of information-users can benefit in their roles as producers, and do so in ways reflected by our traditional, market-oriented indicators of productivity. (David 1990, 360)

The extent to which the engineering (read: business) and political economy professions differ with regard to the effects of automation is best seen in the professional journals. From roughly the 1970's, automation has spawned an extensive literature in engineering, engineering economics, and production economics. Professional journals such as the International Journal of Production Research and the International Journal of Production Economics, to name but two, are replete with work dealing with automation. Titles such as "Automated Productivity Line Design with Flexible Unreliable Machines for Profit Maximization" are indicative of both the purpose and essence of this literature.

Political economy has also spawned a somewhat extensive literature on automation. See Mark Doms, Timothy Dunne and Kenneth Troske's recent article in The Quarterly Journal of Economics for a list of the relevant articles. What is particularly interesting to note about this literature is (i) its insular nature, having no link(s) to the engineering literature, and (ii) its labor-centricity. Both of these, we maintain, can be traced to the underlying model of production. That political economists and engineers today are unable to find common ground is

not surprising given the historical record.   As
such, any meaningful exchange between the two is
beyond reach.   Political economy's labor-centricity,
we maintain, owes to the absence of a model of
production explicitly incorporating information.
Labor acts as an anchor, relating automation to
something political economists are familiar with.
What is ironic is the fact that automation has, is,
and will continue to eliminate labor (lower-level
supervisors) from the workplace, just as inanimate
energy sources eliminated labor in the 19th and 20th
centuries.

As for the move towards freer trade, begun in the
mid 1980's, it too is studied using models which, in
my view, are inappropriate.   The question of
causality is lost in trade theorists' centuries-old
plea, begun by Ricardo, for free trade.   The
globalization of the world economy, characterized by
a shift of manufacturing away from the North and to
the South, is, as such, viewed as an organic
development.   Why corporations which had previously
invested and created hundreds of thousands of jobs
in the North, and who, throughout this century,
opposed free trade with low-wage countries, pushed
for free-trade is a question not even raised.   Trade
is Pareto-improving, period.

As such, trade theorists and trade economists in
general are at a loss when it comes time to
rationalizing the current move towards globalization
and freer world trade. For example, in response to
the question, why did globalization "take off" in
the mid-1980's, the most common response is
developments in information technology.

Conclusions

As we have attempted to show in this chapter, the
20th century will end in much the same way as the 19th
century, namely in a state of energy statis.   As
pointed out in Chapter 2, by the 1890's, the per-
capita level of energy consumption in Great Britain
and the rest of the Western world had reached its
maximum—that is, with the then-existing power
transmission technology, namely shafting, belting,
and gearing.   This, however, is where the
similarities end.   Unlike today where the
possibilities of a new energy form or transmission
technology appear to be limited, the late 19th
century was characterized by important developments
in electro-magnetics, which, as shown in Chapter 3,
ushered in three-quarters of a century of
unparalleled energy deepening, productivity growth,
and rising material well-being.   Such is not the

case at the time of writing. In fact, if anything, the emphasis, since the energy crisis, has been on energy "shallowing" in the form of conservation. There is nothing fundamentally wrong with conservation; however, it must be clearly understood that one cannot have more with less, that one cannot have "growth without growth." That is, society cannot have more output (i.e. work) with less energy. To think otherwise is to violate the basic laws of thermodynamics.

Another important difference is in the area of organization, specifically developments in "organization" technology. As shown in this chapter, developments in inanimate, computer-controlled lower-level forms of supervision have altered and continue to alter the nature of what is commonly referred to as "work." Alfred Marshall's "one woman" managing "four or more looms" has been replaced by a control unit. Clearly, while not yet extinct, it is but a matter of time before animate lower-level supervisors are a thing of the past, only to be found in history books and museums.

5

ICT: The Industrial Revolution That Wasn't

> Without energy, there would be nothing. There
> would be no sun, no wind, no rivers, and no
> life at all. Energy is everywhere, and energy
> changing from one form to another is behind
> everything that happens. Energy, defined as
> the ability to make things happen, cannot be
> created. Nor can it be destroyed. Plants and
> animals harness energy from nature to help them
> grow and survive. The most intelligent of
> animals, human beings, have developed many ways
> of using the available energy to improve their
> lives. Ancient people used energy from fire,
> and they developed tools to use energy from
> their muscles more effectively. But ancient
> people did not understand the role of energy in
> their lives. Such an understanding of energy
> has really developed only over the past few
> hundred years.
>
> —Jack Challoner, Energy

As the three previous chapters make clear,
industrial revolutions have been, at least up until
now, about using more energy to make more goods and
services. James Watt's high-pressure steam engine
and Edison's DC motor paved the way for
unprecedented energy deepening, resulting in
manifold increases in material wealth. It therefore
follows that for a new technology to give rise to an
"industrial revolution," it must have the potential,
ceteris paribus, to increase society's ability to
produce.

This brings us to the so-called information
revolution, or the "third industrial revolution."
More specifically, can and will ICT usher in a third
industrial revolution comparable in magnitude to the
first two? Will it lead to a comparable increase in
the amount of material wealth a society can produce?

Will information and communication do for material
wealth what the BTU's in coal, petroleum and the
atom did over the past two centuries?

This chapter examines these questions critically.
Using the E-O approach to modeling material
processes, it formalizes ICT as an information
consumption enabling technology. ICT, by reducing
the cost of storing and transmitting bytes of
information, enables firms and agents to increase
the rate at which they can produce and consume
information, a concept we refer to as information
deepening. At any given moment, an automated plant
(smart plant) generates more information than a non-
automated one. It will be shown that unlike energy
deepening which increased work and material wealth,
information deepening does not contribute to
increased work and material output owing, in large
measure, to the role played by information in
production processes.

Automation: A Glossary

The following was taken from:
http://www.referenceforbusiness.com/encyclopedia/Ass
em-Braz/Automation.html (public domain).

Automation refers to the use of computers and
other automated machinery for the execution of tasks
that a human laborer (S_a) would otherwise perform.
Companies automate for many reasons. Increased
productivity is normally the primary reason for many
companies desiring a competitive advantage.
Automation can also reduce human error (η) and thus
improve quality. Other reasons to automate include
the presence of a hazardous working environment and
the high cost of human labor. The decision regarding
automation is often associated with some economic
and social considerations.

Virtually every industry sector has benefited
from automation, including manufacturing, services,
and retailing, and some have been greatly
transformed by it. Automation technology falls into
two main categories:

-Physical Process Automation and Control
-Information Management and Processing

Many businesses use both types extensively.

Physical Automation

Physical automation systems are used primarily by
companies that deal in physical products, such as in
mining and manufacturing. Automated machinery may

range from simple sensing devices at one stage of a production process to robots and other sophisticated equipment that control the entire process.

Some of the major classes of physical automation technologies include:

-Computer-Aided Manufacturing (CAM)
-Numerical Control (NC) Equipment
-Robotics
-Flexible Manufacturing Systems (FMS)

Computer-Aided Manufacturing (CIM)

Computer-aided manufacturing (CAM) refers to the use of computers in the different functions of production planning and control. Computer-aided manufacturing includes the use of numerically controlled machines, robots, and other automated systems for the manufacture of products. CAM also includes computer-aided process planning (CAPP), group technology (GT), production scheduling, and manufacturing flow analysis. CAPP means the use of computers to generate process plans for the manufacture of different products. Group technology (GT) is a manufacturing philosophy that aims at grouping different products and creating different manufacturing cells for the manufacture of each group.

Numerical control (NC) machines are programmed versions of machine tools that execute a sequence of operations on parts or products. Individual machines that have their own computers are called computerized numerical control (CNC) machines. When multiple machines share the same computer, they are known as direct numerical control (DNC) machines.

Robots are automated equipment that, through programming, may execute different tasks that are normally handled by a human operator (S_a). In manufacturing, robots are used to handle a number of tasks, including assembly, welding, painting, loading and unloading, inspection and testing, and finishing operations.

Flexible manufacturing systems (FMS) use several kinds of automation to create highly versatile manufacturing processes. They are groups of computer numerical control machines, robots, and an automated material handling system that are used to manufacture a number of similar products or components using different routings among the machines. The alternative routings can provide for rapid product modification in response to market needs or can simply allow a process to keep moving when one machine is out of service. Flexible manufacturing systems have proven to increase manufacturing productivity by 50 percent or more.

A computer-integrated manufacturing (CIM) system is one in which many manufacturing functions are linked through an integrated computer network. These functions can include production planning and control, shop floor control, quality control, computer-aided manufacturing, computer-aided design, purchasing, marketing, and possibly other functions. The objective of a computer integrated manufacturing system is to allow changes in product design, to reduce costs, and to optimize production requirements. In the area of quality control, advanced systems can greatly decrease both human labor (S_a) and the number of defects (η) that go undetected. The most sophisticated of these systems include self-diagnostic functions that alert operators to any processing anomalies (η) and may even be able to fix such problems on their own.

Information Processing Automation

All businesses of any significant size automate their information handling in some way. Because the use of desktop computers is now so common, many simple forms of office automation may be overlooked. Word processing software automates daily tasks such as memo writing and identifying spelling errors. But information management is automated in more powerful ways, as well, such as in electronic identification and tracking of inventory automated record keeping and transaction processing information sharing across the organization data analysis and manipulation

These functions, and many others, can extend to all parts of the business, including finance departments, sales and marketing departments, and even corporate boardrooms. As with physical process automation, information management automation can markedly improve productivity and give corporate management greater strategic control over the enterprise.

Automation's Drawbacks

Despite all the benefits, there are of course problems associated with implementing some kinds of automation. An obvious example is the social and. human costs when automation completely eliminates job categories. While statistics suggest that automation doesn't contribute strongly, if at all, to unemployment on the macroeconomic level, it can lead to personal dislocation and employee resentment. As a result, management must be highly sensitive to these concerns if it wishes to preserve employee morale.

Automation can also fail to deliver on
productivity gains and other intended benefits.
Systems may have technical flaws, or they may have
been designed to emulate an inefficient or overly
complex human process and thus fall short of
enhancing the overall process. Automated systems, in
addition, may have unforeseen negative interactions
with other parts of a process that aren't automated
e.g., if suppliers don't have compatible practices,
a procurement department may not be able to improve
efficiency even if it automates its portion of the
process.

Automation as Seen Through Prism of E-O Approach to Material Processes

This brief résumé of the role automation and
information in material processes can be examined
critically using the E-O approach to modeling
material processes outlined in Chapter 1 and used
extensively in Chapters 2-4. The E-O approach allows
for a better understanding of the exact role of
information and automation in production and sets
the stage for a more thorough examination of the
exact role of information in material processes.

Throughout the resume, we have inserted two
symbols, namely S_a and η. These refer to the role
automation plays in the E-O approach to material
processes. For example, S_a refers to the fact that
automation is seen as a substitute for animate,
human supervision. For example, in the opening
paragraph, the author states:

> Automation refers to the use of computers and
> other automated machinery for the execution of
> tasks that a human laborer (S_a) would otherwise
> perform.

Hence, in this case, automation and information
are substitutes for human, animate supervision,
replacing the human body and more specifically, the
human brain, in overseeing the workings of
production processes.

According to the authors, automation affects
material processes in two fundamental ways, first by
replacing human animate supervision (S_a) and
secondly, by affecting second-law efficiency (η).
Automation, by providing better information and
perhaps more of it, increases second-law efficiency
by (i) reducing down time, (ii) reducing energy
losses, and (iii) reducing feedstock loss. By
providing real time information on such factors as
speed, tension, resistance, operating temperature,
etcetera, automation serves to reduce intervention

lags, feedstock loss, energy loss, etcetera. In so doing, energy efficiency, measured as the ratio of work to energy input, increases. This then brings us to the question of measurement. Is the resulting gain in η significant? What is the order of magnitude of the average gain? Two percent, five percent, ten percent? Is it a one-shot gain or a recurrent one?

Unfortunately, there are no estimates of these gains. However, theoretically speaking, it is reasonable to assume that (i) they are not significant and (ii) they are one-shot gains. The reason is simple, namely that machine breakdown was not a significant problem prior to automation. And, what's more, firms had found ways to deal with it. Hence, while there are advantages, they will be, conceptually speaking, minor.

As for the question of whether such gains will be recurrent, it stands to reason that they will not. Once a plant is "automated," there will be no further gains to be had. In other words, the gains are one-shot in nature, and as such are qualitatively and quantitatively different from those of the first and second industrial revolutions.

Effect on Conventionally-Defined Productivity

While the effects on overall productivity will be minor, the effects on conventionally-defined labor productivity will not, for obvious reasons. Conventionally-defined productivity is defined as overall output divided by the labor input, in this case S_a, animate, human supervision. Automation, by reducing the latter, will increase the ratio despite the presence of negligible output gains. Automation-based labor productivity gains will as such differ qualitatively from first and second industrial revolution-based ones. In the first and second industrial revolutions, energy deepening (greater energy consumption per unit of S_a) increased the numerator. ICT decreases the denominator. For example, in Table 4.2, we see that labor in U.S. manufacturing decreased at an average annual rate of 0.091 percent from 1974 to 1984 when output actually increased. This was also the case in Germany and Japan. This is an important result, one that cannot be overemphasized.

It therefore stands to reason that industries that are more prone to automation will experience greater increases in labor-productivity. As a general rule, the further upstream one goes in the value chain, the greater the capital intensity and the greater the potential for automation. As a corollary, the further down one goes in the value

chain, the lesser is the potential for automation.
For example, automation-related gains in service
sector labor productivity have been minimal as have
been those in the highly labor-intensive industries
of the manufacturing sector. As pointed out in the
previous chapter, the lack of automation-based
profit-increasing opportunities in these sectors has
been one of the driving forces in globalization in
general and outsourcing in particular.

The ICT Revolution Revisited

There can be little doubt: the information
revolution shares little with its first and second
cousins. The latter witnessed massive increases in
energy consumption and consequently in society's
ability to transform raw materials and intermediate
goods into material wealth. The underlying physics
were straightforward: more energy, more work and
more output. As it turns out, this is what is
missing in the so-called third industrial
revolution. The latter was not accompanied by a
massive increase in energy—in fact, it was just the
opposite. As shown in the previous chapter, the
period that corresponds to the third industrial
revolution witnessed a marked reduction in the rate
of growth of energy consumption, ushering in the so-
called productivity slowdown.

This raises the obvious question: if ICT has not
and will not usher in a third industrial revolution
comparable to the first two, then why all the
hoopla? A number of writers have attempted to
address this admittedly difficult question. If ICT
is not output increasing, then what? If information
deepening, unlike the energy deepening of lore, does
not lead to increases in material wealth, then what?

Like Robert Solow, management guru Peter Drucker
remains skeptical as to the effect of ICT on
material processes and material wealth. According to
him, ICT is qualitatively different. The information
revolution is not about more information, but about
the quality of the information in question, and more
importantly, what is actually done with it.

> The next information revolution is well under
> way. But it is not happening where information
> scientists, information executives, and the
> information industry in general are looking for
> it. It is not a revolution in technology,
> machinery, techniques, software, or speed. It
> is a revolution in CONCEPTS.
> So far, for 50 years, the information
> revolution has centered on data their
> collection, storage, transmission, analysis,
> and presentation. It has centered on the "T" in

IT. The next information revolution asks, What is the MEANING of information, and what is its PURPOSE? And this is leading rapidly to redefining the tasks to be done with the help of information, and with it, to redefining the institutions that do these tasks. The next information revolution will surely engulf all major institutions of modern society. But it has started, and has gone farthest, in business enterprise, where it has already had profound impacts. It is forcing us to redefine what business enterprise actually is and should be. This largely underlies the new definition of the function of business enterprise as the "CREATION OF VALUE AND WEALTH," which in turn has triggered the present debate about the "governance of the corporation," that is, for whom the business enterprise creates value and wealth. Yet, despite its importance and impact, the next information revolution has so far been largely ignored by the information establishment. For it has started in the information system of which though it is the oldest and still the most widely used one IT people, as a rule, tend to be both ignorant and contemptuous: Accounting.

This could be interpreted as a veiled admission that ICT is not in fact comparable to previous energy-deepening enabling technologies, but rather, is fundamentally different. ICT will not as such usher in a new era of abundance (quantities) but rather will work primarily on product mix and overall process flexibility.

Computer-aided manufacturing (CAM) refers to the use of computers in the different functions of production planning and control. Computer-aided manufacturing includes the use of numerically controlled machines, robots, and other automated systems for the manufacture of products. CAM also includes computer-aided process planning (CAPP), group technology (GT), production scheduling, and manufacturing flow analysis. CAPP means the use of computers to generate process plans for the manufacture of different products. Group technology (GT) is a manufacturing philosophy that aims at grouping different products and creating different manufacturing cells for the manufacture of each group.

Put differently, ICT improves process planning, group technology, production scheduling, and manufacturing flow analysis, the end result of which is more flexible manufacturing, manufacturing that is more responsive to the needs of consumers and of the market.[1]

This view, however, is not unanimous. Some argue that these gains have been overstated and, more importantly, are limited in scope. For example, Nicholas Carr in an article entitled "IT Doesn't Matter," has argued that most of the transformations that were going to happen as the result of IT, have happened or are in the process of happening.

While the jury is still out on the role of ICT in the reorganization of business enterprise in general, what is clear is the fact that it has not, cannot, and will not usher in a "classical" industrial revolution (i.e. those described in Chapters 2 and 3).

The ICT Revolution as a Continuation of the Second Industrial Revolution

As has been argued elsewhere (Nye 1990), the first and second industrial revolutions were not only about material process-based energy deepening, but were also about the introduction of new enabling technology-based products from the steam locomotive, to the radio, to the refrigerator, to the electric stove, to the vacuum cleaner, etcetera. Common to each of these was an increased use of inanimate energy in the form of either steam or electricity.

It could be argued that the ICT revolution is, in large measure, another by-product, albeit it tardy, of the second industrial revolution. Like all other electronic goods, ICT is electric-energy intensive. Accordingly, it is analogous, in many ways to the radio or the television, the difference being in the way in which information is transmitted and stored. Not surprisingly, the first computers to speak of (e.g. the ENIAC, Univac, Sperry Rand) were based on the same vacuum tube technology that produced the television set.

Multiple ICT Revolutions

As was shown in Chapters 2 and 3, the first and second industrial revolutions can be deconstructed into a number of sub-revolutions, each being defined according to the relevant enabling technology. For example, the first-first industrial revolution was defined in terms of James Watt's condensing, low pressure steam engine. A similar deconstruction can be performed on the ICT revolution. As pointed out, the ICT revolution was about an information-related enabling technology. Early tube-based computers and latter-day transistor/microchip-based computers increased the rate at which individuals could transmit, process and store information.

Table 5.1 presents the various generations of information consumption enabling technologies, measured here by the corresponding CPUs and their speeds, measured in MHz's from 1979 to 1996. We see that the 8088 CPU had a speed of 4.77 MHz, which is dwarfed by the 1996 Pentium Pro which has a speed of up to 200 MHz. Each represents a generation of ICT-enabling technology. Conceptually speaking, with each increase in speed came an increase in the rate at which users could process information (transmit, store, retrieve, etcetera).

Table 5.1
The PC CPU Chart

CPU Type	Year	Speed, MHz	Transitors
8088	1979	4,77	29K
80286	1982	8,10,12,20	130K
80386SX	1988	16,20,25,33	275K
80386SL	1990	16,20,25	855K
80386DX	1985	16,20,25,33	275K
80486SX	1991	16,20,25,33	900K
80486DX	1989	25,33,50	1,200K
80486DX2	1992	25/50,33/66	1,200K
80486SX	1993	25,33	1,400K
80486DX4	1994	25/75,33/100	1,600K
P5 Pentium	1994	75,90,100, 120,133,233, 450,500,800	3,300K
P6 Pentium		133,150,180,	5,500K

Source: www.alcomputers.net/pcpunit.htm

The Future of the ICT Revolution

As we have attempted to show, ICT is an enabling technology, one that has, over the course of the last three decades, increased the rate of "information consumption," both at the individual consumer and producer level. With every increase in machine speed and storage capacity came an increase in information consumption. The question is: what future for the ICT revolution? Where do we go from here?

The answer is twofold. In so far as enabling technology is concerned, further increases in processing/transmission speed are still possible. However, according to a number of experts, there are limits, limits that we are fast approaching. Excerpt

The second issue has to do with the agent's and the firm's consumption of information. Most PC's and laptops have more storage capacity than most users can ever use, at least in a meaningful way. This raises the possibility of information saturation and the distinct possibility of decreasing returns (utility and productivity). "Too much information" is an expression that is finding new meaning in this information-rich era.

Summary

As we have attempted to demonstrate in this chapter, ICT has not, is not, will not and cannot usher in an industrial revolution comparable to the first and second industrial revolutions. The reason is simple: information, unlike energy, is not physically productive. Information-based enabling technologies, while increasing the rate of information consumption, will not increase the rate of growth of output.

The steam engine, the dynamo and ICT all share one important feature, namely they are all enabling technologies per se. Each enables the owner/user to increase the consumption of either energy (BTU's, Joules, kwh's), or information(bytes, KB, MB).

Conclusions

Hope is nature's veil for hiding truth's nakedness.

Alfred Nobel

Economists are, in general, an optimistic bunch. Such was not always the case. Throughout the 19th century and into the 20th century, just the reverse was true. Only with the coming of Keynesian macroeconomics did the tide turn, so to speak. Countries, peoples, classes, individuals were not at the mercy of "the economy" but, rather, could lean into the wind. This optimism characterized most of the 20th century, until of course the productivity slowdown when growth rates throughout the Western world plummeted. Optimism turned into guarded optimism. Surely, the profession that had slain the dragon that was the Great Depression could figure out what had gone wrong, and more importantly, how to fix it.

Underlying this "guarded optimism" was the view that technology shocks powered much of the observed growth in GDP. The steam engine and the dynamo had ushered in decades of prosperity. In the face of the impending ICT revolution (1970s and 1980s), most observers took solace in the fact that while growth rates had in fact gone south, it was just a matter of time before they headed northward again. The expression "Blind Faith" best describes the resulting mood. The profession was blind as to the causes of the productivity slowdown and faithful as to the effects of ICT on growth. Further, just how ICT would increase productivity and growth was left unanswered.

Every ICT innovation, whether it be a faster processor, a new information storage device, or a new medium, raised our subjective priors. Surely, growth could not resist the onslaught of new ICT technology. But it did, it has, and apparently, will continue to. Proverbial cracks began to appear in the hull of the good ship ICT, with the likes of Robert Solow and Robert Gordon firing the first salvos. Given the absence of microfoundations, the debate soon turned into a shouting match with the ICT zealots arguing that "it would increase growth" and the doubting Toms arguing the contrary. This is where our story began.

Clearly, what is lacking (or what was lacking) is a set of solid microfoundations. Shouting matches are not, in general, a good way to resolve what in essence is a scientific debate. Lacking was a scientific model of material processes sufficiently general to incorporate both energy and information and, hence, allow for comparisons of GPT's (General Purpose Technologies), specifically comparisons of the steam engine and ICT, or the dynamo and ICT. In this book, the Energy-Organization approach to modeling material processes was presented and used to examine the first two industrial revolutions and the purported third industrial revolution (ICT Revolution).

The steam engine, the dynamo and the computer (ICT) were formalized as enabling technologies. As energy is the only physically productive factor input, it stands to reason that the steam engine and the dynamo, by contributing (enabling) to further energy deepening, increased productivity and growth. Unfortunately, as information per se is not physically productive, ICT could not, did not, and will not contribute to higher productivity and growth. ICT contributes to (enables) information deepening; however, the latter does not contribute to productivity and output growth.

This is not to say, however, that ICT will not have an effect on wealth. According to the E-O approach to modeling material processes, organization is information intensive. Surely, information deepening will contribute to improvements in second-law efficiency, which in turn will contribute to higher output. The problem, however, is that such gains are punctual in the sense of being one-shot in nature. Second-law efficiency is extremely stable over time, and more importantly, hard to budge.

While some may interpret this finding as being negative, we see it as extremely positive and upbeat. The debate over ICT has served to refocus the debate on fundamentals, fundamentals that shed light not only on ICT but on all GPT's. This is an extremely positive development as policy makers can now focus on growth-related policies armed with a thorough understanding of the growth process.

Endnotes

Introduction

1. Nicholas Crafts (2001) makes a similar argument using the case of steam in early 19th-century Great Britain.
2. For more on the underlying technological specifications, see Dedrick, Gurbaxani and Kraemer (2003), Dale Jorgenson (2001), Brynjolfsson (2003) Atrostic and Nguyen (2005), Simon and Wardrop (2002), Greenan, Mairesse, and Topiol-Bensaid (2001), and Motohashi (2003).

Chapter 1

1. By energy consumption, it should be understood the consumption or use of the available work of a particular energy source, keeping in mind that energy cannot be created or destroyed (First Law of Thermodynamics). As such, energy consumption is equivalent to the concept of entropy.
2. Those interested in the evolution of production processes through time are referred to Chapter 2 of Energy and Organization: Growth and Distribution Reexamined (Westport, Ct: Greenwood Press, 1998).
3. The view that conventionally-defined labor can be broken down into a force (i.e. energy) and supervisory component is as old as thermodynamics itself. German physicist and physiologist Hermann von Helmholtz argued that the forces of nature (mechanical, electrical, chemical, etcetera) are forms of a single, universal energy or Kraft, that cannot be either added to or destroyed. According to Ansom Rabinbach, "As Helmholtz was aware, the breakthrough in thermodynamics had enormous social implications. In his popular lectures and writings, he strikingly portrayed the movements of the planets, the forces of nature, the productive forces of machines, and, of course, human labor power as examples of the principle of conservation of energy. The cosmos was essentially a system of production whose product was the universal Kraft, necessary to power the

engines of nature and society, a vast and protean reservoir of labor power awaiting its conversion to work." (Rabinbach 1990, 3)

4. This has important consequences for distribution. Clearly, the owner(s) of the tools/machines cannot lay claim to a portion of the output on the basis of work. Instead, his/her/their claim has to be based on tools/machines' contribution to second-law efficiency. Tools/machines improve second-law efficiency, thus increasing output.

5. By definition, second-law efficiency consists of the ratio of the minimum theoretical amount of energy required to perform a task, to the actual of amount of energy used in any given production process.

6. I forego a discussion of the notion of mechanical advantage for the simple reason that while tools/machines can better distribute the overall amount of work to be done, they do not, in any way, reduce it. In other words, because they are not a source of energy, they cannot increase the overall amount of work being performed.

7. By spontaneous, it should be understood not having to do with man.

8. One could argue that natural entopic processes are also subject to breakdown. For example, take the human body and the numerous diseases that prevent energy from being transformed into work.

9. Theoretically, this can be expressed as
$\lim_{t \to 0} \eta[s(t), T(t)] = 0$.

10. This explains the increasing use of child labor (i.e. child supervision) in factories in the early 19th century.

Chapter 2

1. In some cases, supervision may have been organized hierarchically, especially where children were involved.

2. Little is known of distribution of income within the family unit. Given that ownership of all the factors was concentrated, the distribution problem was, for all intents and purposes, inexistent.

3. In more practical terms, this implies that every factory worker had the equivalent of 0.682222 of a horse supplying motive power in lieu of human, muscular energy.

4. According to A. E. Musson (1976):
In trying to assess the growth of industrial steam power in this period, one of the main problems is that of measuring "horsepower." Concern for economy in fuel consumption had led in the eighteenth century to production figures on the "duty" of different engines, the unit

being the million ft-lb per bushel (94lb) of coal, for example, by measuring the quantity of water raised, the height through which it was lifted, and the coal consumed by a pumping engine. These figures permitted calculation of the comparative thermal efficiency of engines—greatly improved from Watt's time onwards—and also provided data for calculating From roughly 1841 onwards, increases in horsepower per worker were, in large measure, the result of improvements in the design of and capacity of steam engines.

5.   Conventionally-defined labor productivity is defined as the ratio of output (value added) to labor input.

6. Remember that power looms/spindles operated on a continuous-time basis.

7. For more on the role of energy in the history (archaic and modern) of material civilization, see Chapter 2 of Beaudreau (1998).

8. Moral philosophers, at the time, were, in general, unaware of advances in natural philosophy. Had, for example, Adam Smith been familiar with classical mechanics, as developed by Newton, Leibnitz, and Laplace, political economy may have taken a different turn. It is my view that had Smith, and to a lesser degree the French Physiocrats, adopted natural philosophy as their underlying model, then political economy would have been spared two centuries of missed opportunities, paradoxes, and, failures.

9. Remark that we do not maintain that the Wealth of Nations is the first attempt at understanding wealth in general. As is well known, the Physiocrats had set out to study wealth from a scientific point of view as early as 1759. Given Smith's familiarity with this body of literature, it could be argued that the Wealth of Nations is synthetic in nature, consisting of Physiocracy augmented by water- and steam-powered machinery. In passing, it should be noted that Physiocracy is consistent with the analytical framework presented in Chapter 1. According to the Physiocrats, agriculture is the only source of "economic surplus." Another way of seeing this is in terms of energy rents. Specifically, only in agriculture were energy rents, defined as the difference between output and costs, possible.

10. From a strictly physical point of view, the only way in which labor can be more productive is for it, as a source of energy, to use better tools (higher second-law efficiency), or for it to exert more energy/force.

11. Smith's hunting metaphor attests to the Paleolithic nature of classical value theory.
12. This has since become known as the "trickle down effect."
13. One could argue that, by now, its productivity was marginal.
14. Judging from the content of the The Coal Question, it is clear that Jevons was, prior to Das Kapital, well on his way to an energy theory of value. Expressions such as "As the source of fire, it is the source of mechanical motion and of chemical change," convey the import he attributes to energy in general and coal in particular.
15. Interestingly enough, Philip Mirowski, in More Heat than Light, Economics as Social Physics, Physics as Nature's Economics, does not address this question, preferring to focus, for the most part, on the use by neoclassical political economists of the field metaphor to describe utility and production.
16. This way of modeling energy shocks is very much alive and well in the literature today. Take, for example, Timothy Bresnahan, Elhanan Helpman and Manuel Trajtenberg's work on "General Purpose Technologies," where energy shocks are viewed as one-shot phenomena (Bresnahan and Trajtenberg 1992; Helpman and Trajtenberg 1994). Specifically, "GPT (General Purpose Technologies) innovations such as the "steam engine, electricity or micro-electronics" prompt investments in complementary inputs, thereby giving rise to "recurrent cycles."
During the first phase of each cycle output and productivity grow slowly or even decline, and it is only in the second phase that growth starts in earnest. The historical record of productivity growth associated with electrification, and perhaps also of computerization lately, may offer supportive evidence for this pattern. (Helpman and Trajtenberg 1994)

Chapter 3

1. One could argue that muscular energy is also a power transmission technology, converting the the chemical energy (adenosine triphospates) found in carbohydrates and proteins into work (movement).
2. Nuclear power in this period is relatively insignificant. For example, in 1956, 10 million kWh's of the net generation of electricity was from nuclear sources, which corresponds to .00209 of total net generation.
3. Ford's Highland Park plant, one could argue, marks the crossover from high-throughput

continuous-flow production processes powered by
either water or steam power, to extremely-high-
throughput continuous-flow production processes
powered by electricity. Put differently, the
looms and spinning jennies of the early 19th
century, powered by steam power, constituted
high-throughput, continuous-flow production
processes. The electric-powered stamping
machines, the electric-powered material handling
systems, the electric powered assembly lines, on
the other hand, constituted extremely-high (in a
comparative sense), continuous-flow production
processes.
4. It need be pointed out that not all of this
represents a net increase in energy consumption
as electric drive had not completely displaced
steam drive in industry.
5. Perhaps this explains the dominance of the U.K.
in 19th century political economy and the near
absence of Continental and North American
political economy.
6. Chief among the reasons for choosing the direct
method was the purported absence of competitive
factor markets (energy and labor) (Blanchflower,
Sanfey and Oswald 1996; Van Reenen 1996).
Specifically, the electric power market is
regulated, via public utility commissions, while
the labor market, according to Balnchflower et
al. and Van Reenen, was found to be non-
competitive.

Chapter 5

1. For more on this, see Greenmeier (2005),
Hidebrand (2005), Whiting (2003), and Reddy
(2004).

Bibliography

Alterman, Jack. A Historical Perspective on Changes
in U.S. Energy-Output Ratios. Bulletin EA-3997
Palo Alto, CA: Electric Power Research Institute
1985.

Alting, Leo. Manufacturing Engineering Processes.
New York, NY: Marcel Decker Inc., 1994.

Anderson, F.J., N.C. Bonsor and B.C. Beaudreau. The
Economic Future of the Forest Products Industry in
Northern Ontario. Royal Commission on the Northern
Environment. Thunder Bay, ON 1982.

Atrostic, B. K., and Sang V. Nguyen. "IT and
Productivity in U.S. Manufacturing: Do Computer
Networks Matter?" Economic Inquiry 43.3 (July 2005):
493-506.

Ayres, Robert U. and Nair, Indira. Thermodynamics
and Economics. Physics Today 1984 62-71.

Babbage, Charles. The Economy of Machinery and
Manufacturing. London: C. Knight, 1832.

Baily, Martin N. and Gordon, Robert J. The
Productivity Slowdown. Measurement Issues. and the
Explosion of Computer Power. Brookings Papers on
Economic Activity 2 1988 347-420.

Baines, E. The History of the Cotton. Manufactures
London, 1835.

Barro, Robert J. and Sala-i-Martin. Xavier. Economic
Growth. New York. NY: McGraw-Hill, 1995.

Beaudreau, Bernard C. The Impact of Electric Power
on Productivity: The Case of U.S. Manufacturing
1958-1984. Energy Economics 17(3) 1995a 231-236.

Beaudreau, Bernard C. The Impact of Electric Power
on Productivity: The Case of Japanese Manufacturing

1965-1988. Working Paper Department of Economics Université Laval 1995b.

Beaudreau, Bernard C. The Impact of Electric Power on Productivity: The Case of German Manufacturing 1963-1988. Working Paper. Department of Economics Université Laval 1995c.

Beaudreau, Bernard C. Newtonian Production Processes. Working Paper. Department of Economics Université Laval 1995d.

Beaudreau, Bernard C. Mass Production, The Stock Market Crash, and The Great Depression: The Macroeconomics of Electrification Westport. CT: Greenwood Press, 1996a.

Beaudreau. Bernard C. R D: To Compete or to Cooperate. Economics of Innovation and New Technology. 4 1996b 173-186.

Beaudreau. Bernard C. Energy and Organization: Growth and Distribution Reexamined. Westport. CT: Greenwood Press, 1998.

Beaudreau, Bernard C. Energy and the Rise and Fall of Political Economy. Westport, CT: Greewood Press, 1999.

Beiser, Arthur. Modern Technical Physics. Menlo Park, CA: The Benjamin/Cummings Publishing Company, 1983.

Berg. Maxine. The Machinery Question and the Making of Political Economy. Cambridge: Cambridge University Press, 1980.

Berndt, Ernst and David O. Wood. Technology. Prices and the Derived Demand for Energy. The Review of Economics and Statistics. August 1975 259-268.

Betts, John E. Essentials of Applied Physics. Englewood Cliffs. NJ: Prentice-Hall, 1989.

Birnie, Arthur. An Economic History of Europe 1760-1930. New York. NY: The Dial Press, 1930.

Blair, John M. Does Large-Scale Eneterprise Result in Lower Costs? American Economic Review 38 1948 121-152.

Blanchflower, David G.,Oswald. Andrew J.. and Sanfey. Peter. Wages. Profits. and Rent Sharing. Quarterly Journal of Economics 60(1) 1996 227-251.

Bose, Bismal K. Introduction to Microcomputer Control. in Bose, Bismal K. ed.. Microcomputer Control of Power Electronics and Drives. New York, NY: IEEE Press, 1987 .

Bresnahan, Timothy and Manuel Trajtenberg. General Purpose Technologies: Engines of Growth? National Bureau of Economic Research Working Paper No. 4148. August 1994.

Butt, John. Robert Owen: Prince of Cotton.   Newton Abbott: David and Charles, 1971.

Casson, Lionel. Ancient Trade and Society.   Detroit, MI: Wayne State University Press. 1984 .

Chase, Stuart. The Economy of Abundance.   New York, NY: MacMillan Company, 1934.

Challoner,   Jack.   Energy.   London: Darling Kindersley, 1993.

Chandler, Alfred D. Jr. The Visible Hand.   The Managerial   Revolution   in   American   Business. Cambridge, MA: Harvard University Press, 1977.

Crafts, Nicholas. Forging Ahead and Falling Behind: The   Rise   and   Relative   Decline   of   the   First Industrial Nation. Journal of Economic Perspectives 12(2) 193-210.

Christensen, L. R.. and D.W. Jorgenson. U.S. Real Product and Real Factor Input. The Review of Income and Wealth 1970.

Clapham, J.H. Of Empty Economic Boxes. Economic Journal. 32 1922 305-314.

Clark, J. Maurice. Studies in the Economics of Overhead Costs.   Chicago, IL: University of Chicago Press, 1923.

Cleveland, Cutler J. Costanza, R. Hall, Charles. and Kaufmann, R. Energy and the U.S. Economy: A Biophysical Perspective. Science 225 1984 890-897.

Clower, Robert and Peter Howitt. Keynes and the Classics: An End of Century View. in Ahiakpor. James C.W.   ed.   Keynes and the Classics Reconsidered. Boston, MA: Kluwer, 1998.

Clower,   Robert.   A   Reconsideration   of   the Microfoundations   of   Monetary   Theory.   Western Economic Journal. 1967 1-19.

Cobb, Charles and Douglas, Paul A Theory of Production. American Economic Review. 18 1928 139-165.

Coleman, Matthew J. ed. Energy Engineering and Management in the Pulp and Paper Industry. Atlanta, GA: TAPPI, 1991.

Copithorne, Lawrence W. A Neoclassical Perspective on Natural Resource-Led Regional Economic Growth. Economic Council of Canada Discussion Paper no. 92 1977.

David, Paul A. The Dynamo and the Computer: An Historical Perspective on the Modern Productivity Paradox. American Economic Review Papers and Proceedings. May 1990 355-361.

Dean, Joel. Cost-Structures of Enterprises and Break-Even Charts. American Economic Review 38 1948 153-164.

Dedrick, Jason, ViJay Gurbaxani, and Kenneth L. Kraemer. "Information Technology and Economic Performance: A Critical Review of the Empirical Evidence." *ACM Computing Surveys* 35.1 (Mar. 2003): 1-28.

Demsetz, Harold. Industry Structure. Market Rivalry. and Public Policy. Journal of Law and Economics 16 1-9.

Demsetz, Harold. Two Systems of Belief about Monopoly. in Goldschmid et al. eds. Industrial Concentration. The New Learning. Boston, MA: Little Brown, 1974.

Denison, Edward F. The Sources of Economic Growth in the United States and the Alternatives Before Us. New York, NY: Committee for Economic Development, 1962.

Denison, Edward F. Trends in American Economic Growth 1929-1982. Washington, DC: The Brookings Institution, 1985.

Devine, Warren D. Electricity in Information Management: The Evolution of Electronic Control. in Schurr. Sam. H. et al. eds. Electricity in the American Economy. Westport ,CT: Greenwood Press, 1990 .

Dimand, Robert W. Cranks. Heretics and Macroeconomics in the 1930's. History of Economic Review 16 1991 11-30.

Doms, Mark, Dunne, Timothy. and Troske, Kenneth R. Workers, Wages, and Technology. Quarterly Journal of Economics 112 253-289.

Douglas, Clifford H. Social Credit. New York, NY: W.W. Norton and Company, 1933.

Douglas, Clifford H. The Monopoly of Credit. Liverpool: K.R.P. Publications, 1951.

Du Boff, Richard B. The Introduction of Electric Power in American Manufacturing. Economic History Review 1967 509-518.

Du Boff, Richard B. Electrical Power in American Manufacturing 1889-1958. New York. NY: Arno Press, 1979.

Edgeworth, Francis Y. Mathematical Psychics. London: Kegan Paul, 1881.

Fielden, John. National Regeneration. in Carpenter, Kenneth E. ed. The Factory Act of 1833. New York. NY: Arno Press, 1972.

Fischer, Stanley. Money and the Production Function. Economic Inquiry 12 1974 517-533.

Fisher, Irving. The Purchasing Power of Money. New York. NY: MacMillan, 1911.

Floud, Roderick and McCloskey, Donald. eds. The Economic History of Britain Since 1700 Volume 1: 1700-1860. Cambridge: Cambridge University Press, 1994.

Ford, Henry. Mass Production. Encyclopaedia Britannica 13 1926 821-823.

Friedman, Milton and Anna J. Schwartz. A Monetary History of the United States 1867-1960. New York: National Bureau of Economic Research, 1963.

Galbraith, John K. American Capitalism: The Concept of Countervailing Power . New Brunswick, NJ: Transaction Publishers, 1952 -1993.

Georgescu-Roegen, Nicholas. The Entropy Law and the Economic Process. Cambridge, MA: Harvard University Press, 1971.

Gollop, F.M. and Jorgenson, D.W. U.S. Productivity Growth by Industry 1948-1973. in Kendrick, J.W.. and Vaccara, B.N. eds. New Developments in

Productivity Measurement and Analysis. Chicago, IL: National Bureau of Economic Research, 1980.

Gordon, Robert J. Supply Shocks and Monetary Policy Revisited. American Economic Review 74(2) 1984 38-43.

Gorgel, John Francis. A Theory of the Military-Industrial Firm. in Seymour Melman. ed. The War Economy of the United States. New York: St. Martin's Press, 1971.

Graham, Frank D. Relation of Wage Rates to the Use of Machinery. American Economic Review. 16(3) 1926 434-442.

Grant, Michael and Rachel Kitzinger. eds. Civilization of the Ancient Mediterranean. Greece and Rome. New York: Charles Scribner's Sons, 1988.

Greenan, N, J. Mairesse and A. Topiol-Bensaid. "IT and Research and Development Impacts on Productivity and Skills: Looking for Correlations on French Firm Level Data." in IT: Productivity, and Economic Growth. Ed. M Pohjola. New Delhi: Oxford University Press, 2001: 119-148.

Greenemeier, Larry. "Better Data Makes A Powerful Potion." InformationWeek 19 Sep. 2005 <www.informationweek.com/shared/printableArticle.jht ml?articleID=170703697>.

Greider, William. One World Ready or Not, The Maniac Logic of Global Capitalism. New York: Simon and Schuster, 1997.

Griliches, Zvi. The Discovery of the Residual: An Historical Note. National Bureau of Economic Research Working Paper 5348 1995.

Grossman, Gene. and Elhanan Helpman. Innovation and Growth in the Global Economy. Cambridge, MA: The MIT Press, 1991.

Gullickson William and Michael J. Harper. Multifactor Productivity in U.S. Manufacturing 1949-1983. Monthly Labor Review 1988 18-28.

Harley, C. Knick. Reassessing the Industrial Revolution: A Macro View. in Moykr, Joel. Ed. The British Industrial Revolution: An Economic Perspective. Oxford: Westview Press, 1993 171-226.

Hayes, H. Gordon. Rate of Wages and the Use of Machinery. American Economic Review 13(3) 1923 461-465.

Helpman, Elhanan. ed. General Purpose Technologies and Economic Growth. Cambridge, MA: MIT Press, 1998.

Helpman, Elhanan and Manuel Trajtenberg. A Time to Sow and a Time to Reap: Growth Based on General Purpose Technologies. National Bureau of Economic Research Working Paper No. 4854 September 1994.

Hildebrand, Carol. "Electric Money." Profit Magazine Feb. 2005.

Hills, Richard L. Power in the Industrial Revolution. Manchester: Manchester University Press, 1970.

Hills, Richard L. Power from Steam: A History of the Stationary Steam Engine . Cambridge: Cambridge University Press, 1989.

Hisnanick, John J. and Kymm, Kern. The Impact of Disaggregated Energy on Productivity. Energy Economics 1992 274-278.

Hollander, Samuel. The Economics of Adam Smith. Toronto: University of Toronto Press, 1973.

Hollander, Samuel. Classical Economics. New York. NY: Basil Blackwell, 1987.

Honeyman, Katrina. Origins of Entreprise: Business Leadership in the Industrial Revolution. Manchester: Manchester University Press, 1982.

Hounshell, David A. From the American System to Mass Production 1800-1932: The Development of Manufacturing Technology in the United States. Baltimore, MD: The Johns Hopkins University Press, 1984.

Hulten, Charles R. Growth Accounting When Technical Change is Embodied in Capital. American Economic Review September 1992 964-980.

Iinglis, Brian. Men of Conscience. New York: The Macmillan Company, 1971.

Jensen, M.C. and Murphy. K.J. Performance Pay and Top-Management Incentives. Journal of Political Economy 98 1990 225-264.

Jevons, W.S. The Coal Question. London: MacMillan and Co., 1865.

Jevons, W.S. The Theory of Political Economy. London: Pelican Books, 1871.

Johnson, H.G. The Efficiency and Welfare Implications of International Corporations. in Kindleberger, C.P. ed. The International Corporation. Cambridge, MA: MIT Press, 1970.

Jorgenson, D.W. Energy Prices and Productivity Growth. in Schurr, S. et al. eds. Energy. Productivity and Economic Growth. Cambridge, MA: Oelgeschlager. Gunn, and Hain, 1983.

Jorgenson, D.W. The Role of Energy in Productivity Growth. in Kendrick. J.W. ed. International Comparisons of Productivity and Causes of the Slowdown. Cambridge, MA: MIT Press, 1981.

Jorgenson, D.W. and B. Fraumeni. Relative Prices and Technical Change. in Berndt, E.R. and B. Field eds. Modeling and Measuring Natural Resource Substitution. Cambridge, MA: MIT Press,1981.

Jorgenson, Dale W. "Information Technology and the U.S. Economy." American Economic Review 91.1 (Mar. 2001): 1-32.

Kahn. Alfred E. The Economics of Regulation: Principles and Institutions. Cambridge, MA: MIT Press, 1988.

Keynes, John Maynard. Letter to President Roosevelt. New York Times. December 31. 1933 reproduced in Moggridge. Donald. ed.. The Collected Writings of John Maynard Keynes Volume XXI. Cambridge: Cambridge University Press .

Keynes, John Maynard. The General Theory of Employment, Interest and Money London: Macmillan, 1936.

Knight, F. H. Risk, Uncertainty and Profit Chicago, IL: University of Chicago Press, (1921) 1971.

Knight, K. G. Unemployment: An Economic Analysis. London: Croom Helm, 1987.

Lacey, Robert. Ford: The Men and the Machine. Boston, MA: Little. Brown and Company, 1986.

Laidler, David. The Demand for Money: Theories and Evidence. New York, NY: Dun-Donnelley, 1977.

Laidler, David. Taking Money Seriously and Other
Essays. Cambridge, MA: MIT Press, 1990.

Lehr, William and Frank R. Lichtenberg. Computer Use
and Productivity Growth in Federal Government
Agencies. National Bureau of Economic Research.
Working Paper 5616 1996.

Leibenstein. Harvey. Entrepreneurship and
Development. American Economic Review 58 72-83.

Lippman. S.A. and J.J. McCall. The Economics of Job
Search. Economic Inquiry.14 1976 155-189.

Lloyd George, David. Coal and Power London: Hodder
and Stoughton, 1924.

Lutz, Frederich. Corporate Cash Balances 1914-1943.
New York, NY: National Bureau of Economic Research,
1945.

Maddison, Angus. Growth and Slowdown in Advanced
Capitalist Economies: Techniques of Quantitative
Assessment. Journal of Economic Literature 25 2
649-698.

Malthus, T.R. Principles of Political Economy
Considered with a View to Their Practical
Application. New York, NY: Augustus M. Kelley,
1951.

Mankiw, N. G. and D. Romer. New Keynesian Economics.
Cambridge, MA: MIT Press, 1991.

Mannisto. Heikki. Who can afford to save energy? in
Coleman. Matthew J. ed.. Energy Engineering and
Management in the Pulp and Paper Industry. Atlanta,
GA: TAPPI, 1991.

Mansfield, E.. J. Rapoport, J. Schnee, S. Wagner and
M. Hamburger. Research and Innovation in the Modern
Corporation. New York, NY: W.W. Norton and Co.,
1971.

Marshall, Alfred. Principles of Economics. 8th Ed.
London: MacMillan, 1890.

Marx, Karl. Das Capital. Chicago, IL: Encyclopaedia
Britannica (1867) 1992.

Mathias, Peter. The First Industrial Nation: An
Economic History of Britain 1700-1914. London:
Methusen Co., 1969.

McCloskey, Donald. 1780-1860: A Survey. in Floud. Roderick and McCloskey. Donald.   eds.   The Economic History of Great Britain Since 1700. Volume 1: 1700-1860. Cambridge: Cambridge University Press, 1994.

Means, Gardiner. The Growth in the Realtive Importance of the Large Corporation in American Economic Life. American Economic Review 21 1931 10-42.

Merill, Milton R. Reed Smoot: Apostle in Politics. Logan, UT: Utah State Press, 1990.

Mirowski, Philip. Energy and Energetics in Economic Theory: A Review Essay. Journal of Economic Issues. 22(3) 1988 811-830.

Mirowski, Philip. More Heat Than Light, Economics as Social   Physics,   Physics   as   Nature's   Economics. Cambridge: Cambridge University Press, 1989.

Mitchell,   B.R.   British   Historical   Statistics. Cambridge: Cambridge University Press,1988.

Mokyr, Joel.   Technological Change. 1700-1830. in Floud. Roderick and McCloskey. Donald.   eds.   The Economic History of Great Britain Since 1700. Volume 1: 1700-1860. Cambridge: Cambridge University Press, 1994.

Motohashi, Kazuyuki. "Firm Level Analysis of Information Network Use and Productivity in Japan." Paper Presented at the Comparative Analysis of Enterprise Data Conference Sep. 2003.

Moulton, Harold G. Financial Organization and the Economic System. New York, NY: McGraw-Hill, 1938.

Mowery, D.C. and N. Rosenberg. Technology and the Pursuit of Economic   Growth.   Cambridge: Cambridge University Press, 1989.

Musson, A. E. Industrial Motive Power in the United Kingdom 1800-70. Economic History Review. 29 1976 415-439.

Newcomb, Simon. Principles of Economics.   New York, NY: Augustus Kelley, 1886.

Nonneman, Walter and Vanhoudt, Patrick. A Further Augmentation of the Solow Model and the Empirics of Economic Growth for OECD Countries. Quarterly Journal of Economics 1997 112 943-953.

Nye, David E. Electrifying America: Social Meaning of a New Technology Cambridge, MA: MIT Press, 1990.

Odum, H.T. and E.C. Odum. Energy Basis for Man and Nature. New York, NY: McGraw-Hill, 1976.

Owen, Robert. The Life of Robert Owen. London: Cass, 1967.

Owen, Robert. A New View of Society and Other Writings. London: J.M. Dent and Sons. Ltd., 1927.

Patinkin, Don. Money, Interest and Prices. New York, NY: Harper and Row, 1965.

Pigou, Arthur C. The Value of Money. Quarterly Journal of Economics. 37 4 38-65.

Prescott, Edward C. Needed: A Theory of Total Factor Productivity. International Economic Review 39(3) 1998.

Pullen, J.M. and G.O. Smith. Major Douglas and Social Credit: A Reappraisal. History of Political Economy 29 1997 219-273.

Rabinbach, Anson. The Human Motor: Energy, Fatigue and the Origins of Modernity. New York, NY: Basic Books, 1990.

Reddy, P. Krishna. "A Framework of Information Technology Based

Ricardo, David. The Principles of Political Economy and Taxation. New York, NY: Everyman's Library, 1965.

Rifkin, Jeremy. The End of Work. New York, NY: G.P. Putnam's Sons, 1995.

Romer, Paul M. Crazy Explanations for the Productivity Slowdown. NBER Macroeconomics Annual 1987. 1987 163-202.

Romer, Paul M. Endogenous Technological Change. Journal of Political Economy 98 1990 s71-s102.

Romer, Paul M. Increasing Returns and Long-Run Growth. Journal of Political Economy 94 1986 1002-1037.

Romer, Paul M. The Origins of Endogenous Growth. Journal of Economic Perspectives 8 1994 3-22.

Rosenberg, Nathan. Technology and American Economic Growth. Armonk, NY: M.E. Sharpe, 1972.

Rosenberg, Nathan. The Effects of Energy Supply Characteristics on Technology and Economic Growth. in Schurr. S. et al. eds. Energy. Productivity and Economic Growth. Cambridge, MA: Oelgeschager. Gunn and Hain, 1983.

Rosenberg, Nathan and L.E. Birdzell. Jr.. How the West Grew Rich: The Transformation of the Industrial World. New York, NY: Basic Books, 1986.

Rotemberg, Julio and Michael Woodford. Imperfect Competition and the Effects of Energy Price Increases on Economic Activity. National Bureau of Economic Research Working Paper 5634 1996.

Ruth, Matthias. Integrating Economics, Ecology and Thermodynamics. Dordrecht: Kluwer Academic Publishers, 1993.

Ruth, Matthias and Clark W. Bullard. Information. Production and Utility. Energy Policy October 1993 1059-1067.

Schurr, S.H. and Netschert. B. Energy in the American Economy. 1850-1975 Baltimore. MD: Johns Hopkins, 1960.

Schurr, S., J. Darmstadler, H. Perry, W. Ramsay and M. Russell. Energy in America's Future. Baltimore, MD: Johns Hopkins Press, 1979.

Schurr, S., Burwell, C., Devine, W., and Sonenblum, S., Electricity in the American Economy: Agents of Technological Progress Westport. CT: Greenwood Press, 1990.

Scott, Howard et al. Introduction to Technocracy. New York, NY: The John Day Company, 1933.

Self, Sir Henry and Elizabeth M. Watson. Electricity Supply in Great Britain. London: Allen and Unwin, 1952.

Simon, Herbert. The Sciences of the Artificial Cambridge, MA: MIT Press, 1996.

Simon, John and Sharon Wardrop. "Australian Use of IT and its Contribution to Growth." Research Discussion Paper. Economic Research Department, Reserve Bank of Australia (Jan. 2002).

Sinclair, Upton. The Autobiography of Upton Sinclair. New York, NY: Harcourt. Brace and World, 1962.

Smith, Adam. An Inquiry into the Nature and Causes of the Wealth of Nations.   Chicago: Encyclopaedia Britannica, (1776) 1990.

Sobel, Robert. The Age of Giant Corporations, A Mciroeconomic History of American Business 1914-1970. Westport, CT: Greenwood Press, 1972.

Soddy, Frederick. Cartesian Economics: The Bearing of Physical Sciences upon State Stewardship. London: Hendersons, 1924.

Solow, Robert M. The Economics of Resources or the Resources of Economics. American Economic Review 64(2) 1974 1-14.

Solow. Robert M.. Perspectives on Growth Theory. Journal of Economic Perspectives 8 1994 45-54.

Sowell, Thomas. Say's Law: An Historical Analysis. Princeton, NJ: Princeton University Press, 1972.

Spreng, Daniel T.. Possibilities for Substitution between Energy, Time, and Information. Energy Policy 1993 13-23.

Stearns, Peter. The Industrial Revolution in World History. Boulder, CO: Westview Press, 1993.

Stern, David I. Energy and Economic Growth in the USA. Energy Economics. April 1993 137-151.

Stocking, Collis A. Modern Advertising and Economic Theory. American Economic Review 21 1931 43-55.

Tatom, J. Potential Output and the Recent Productivity Decline.   Economic Review of the Federal Reserve Bank of St-Louis 64 1982 3-15.

Temin, Peter. Two Views of the British Industrial Revolution. Journal of Economic History March 1997.

Thomas, Woodlief. The Economic Significance of the Increased Efficiency of American Industry. American Economic Review 18 1928 122-138.

Tugwell, Rexford G. Industry's Coming of Age.   New York, NY: Columbia University Press, 1927.

Tylecote, Mabel. The Mechanics Institutes of Lancashire and Yorkshire before 1851. Manchester: Manchester University Press, 1957.

Tyron, F.G. An Index of Consumption of Fuels and Water Power. Journal of the American Statistical Association 22 1927 271-282.

United Nations. Industrial Statistics Yearbook 1984. New York, NY: United Nations, 1960-1988.

U.S. Department of Commerce. Historical Statistics of the U.S.: Colonial Times to 1970. Bicentennial Edition. Washington, DC: Bureau of the Census, 1975.

U.S. Department of Commerce. Annual Survey of Manufactures. Washington, DC: Bureau of the Census, various years.

U.S. Department of Commerce. Survey of Current Business. Washington, DC: Bureau of Economic Analysis, January 1986.

Van Reenen, John. The Creation and Capture of Rents: Wages and Innovation in a Panel of U.K. Companies. Quarterly Journal of Economics. 61(1) 1996 195-226.

Varian, Hal R. Microeconomic Analysis. New York, NY: Norton, 1992.

Veblen, Thornstein. The Engineers and the Price System. New York, NY: Augustus M. Kelley, 1965.

von Tunzelmann, G. N. Steam Power and Bristish Industrialization to 1860. Oxford: Clarendon Press, 1978.

Wicksell, Knut. Interest and Prices. A Study of the Causes Regulating the Value of Money. Macmillan and Co.: London, 1936 .

Whitaker, J.K. The Early Writings of Alfred Marshall. 1867-1890. New York, NY: The Free Press, 1975.

White, Horace G. A Review of Monopolistic and Imperfect Competition Theories. American Economic Review 26 1936 637-649.

Whiting, Rick. "The Data-Warehouse Advantage." InformationWeek 28 Jul. 2003. <www.informationweek.com/shared/printableArticle.jht ml?articleID=12802974>.

Woolf, Arthur G. Electricity. Productivity and Labor Saving: American Manufacturing 1900-1929. Explorations in Economic History 21 1984 176-191.

Woirol, Gregory R. The Technological Unemployment and Structural Unemployment Debates. Westport, CT: Greenwood Press, 1996.

Wright, Gavin. The Origins of American Industrial Success 1879-1940. American Economic Review 80(4) 1990 651-668.